T0312755

60 GHz TECHNOLOGY FOR GBPS WLAN AND WPAN

60 GHz TECHNOLOGY FOR GBPS WLAN AND WPAN

FROM THEORY TO PRACTICE

Su-Khiong (SK) Yong
Marvell Semiconductor, USA

Pengfei Xia
Broadcom Corporation, USA

Alberto Valdes-Garcia
IBM, USA

⊕WILEY

A John Wiley and Sons, Ltd., Publication

This edition first published 2011
© 2011 John Wiley & Sons Ltd.

Registered office
John Wiley & Sons Ltd, The Atrium, Southern Gate, Chichester, West Sussex, PO19 8SQ, United Kingdom

For details of our global editorial offices, for customer services and for information about how to apply for permission to reuse the copyright material in this book please see our website at www.wiley.com.

Library of Congress Cataloging-in-Publication Data

Yong, Su-Khiong.
  60 GHz technology for Gbps WLAN and WPAN : from theory to practice /
Su-Khiong Yong, Pengfei Xia, Alberto Valdes Garcia.
       p. cm.
  Includes bibliography and index.
  ISBN 978-0-470-74770-4 (cloth)
 1. Millimeter wave communication systems. 2. Wireless LANs. 3. Gigabit communications.
4. Wireless communication systems. I. Xia, Pengfei. II. Garcia, Alberto Valdes. III. Title.
  TK5103.4835.Y66 2011
  621.384–dc22

                                                    2010022098

A catalogue record for this book is available from the British Library.

Print ISBN 9780470747704 (H/B)
ePDF ISBN: 9780470972939
oBook ISBN: 9780470972946

Typeset in 11/13pt Times-Roman by Laserwords Private Limited, Chennai, India
Print and Bound in Singapore by Markono Print Media Pte Ltd

To my wonderful family Chia-Chin, Jerrick and my parents
– Su-Khiong (SK) Yong

To Dad, Mom, Wenjun and Niuniu
– Pengfei Xia

To my daughter Cecilia who came into existence
along with this book
– Alberto Valdes-Garcia

Contents

Preface

Since the first wireless transatlantic radio wave transmission (based on long wave) by Marconi from Cornwall, England, to Newfoundland, Canada, in 1901, wireless communications have undergone tremendous growth. Today, wireless communications systems have become an integral part of our daily life and continue to evolve in providing better quality and user experience.

One of the most important emerging wireless technologies in recent years is millimeter-wave (mm-wave) technology. Although it has been known for many decades, it is only over the past five or six years that advances in silicon process technologies and low-cost integration solutions have made mm-wave a relevant technology from a commercial perspective. As a result, this technology has attracted significant interest from academia, industry and standardization bodies. In this book, we specifically focus on 60 GHz wireless systems that enable several new applications that are not feasible at lower carrier frequencies.

60 GHz technology offers various advantages over current or existing communications systems. One of the most important is the availability of at least 5 GHz of continuous bandwidth worldwide. While this is comparable to the unlicensed bandwidth allocated for ultra-wideband (UWB) purposes, the 60 GHz bandwidth is continuous and less restricted in terms of power limits. In fact, the large bandwidth at 60 GHz band is one of the largest unlicensed bandwidths being allocated in history. This huge bandwidth represents great potential in terms of capacity and flexibility, making 60 GHz technology particularly attractive for gigabit wireless applications. The compact size of the 60 GHz radio also permits multiple-antenna solutions at the user terminal that are otherwise difficult if not impossible at lower frequencies. Compared to 5 GHz system, the form factor of mm-wave systems is approximately 140 times smaller and thus can be conveniently integrated into consumer electronic products.

Despite the various advantages offered, mm-wave based communications face a number of important challenges that must be solved. This book outlines the challenges, opportunities and current solutions at every layer of a 60 GHz system implementation. The outline of the book is as follows.

Chapter 1 presents an introduction to 60 GHz technology. It starts with direct comparisons between 60 GHz technology and other high data rate counterparts such as

UWB and IEEE 802.11n technologies in terms of their transmit power, bandwidth and spectrum efficiency in delivering high data rate solutions. Several key applications that have proved challenging in the past become feasible with the Gbps data rate of 60 GHz. The worldwide regulatory and frequency allocation of the 60 GHz band is then introduced. Finally, intensive standardization efforts are discussed and a comparison of their physical layer features is provided.

Chapter 2 presents an overview of 60 GHz channel modeling, which forms the basis for reliable 60 GHz wireless communications system design. This chapter begins by highlighting the different modeling approaches available and setting out their advantages and disadvantages in generating a realistic 60 GHz channel model. Next, generic modeling frameworks for both large- and small-scale channel characterizations are thoroughly discussed. An extensive list of references with summary and comprehensive discussions on the reported results is provided. This chapter also discusses 60 GHz polarization modeling methodology for multi-polarized multiple-antenna systems. Finally, channel parameterizations for the proposed generic channel models are provided. In particular, the channel models used in IEEE 802.15.3c and IEEE 802.11.ad are discussed and their limitations also highlighted.

Chapter 3 describes radio frequency (RF) nonlinearities and their behavioral models which should be considered in the design of 60 GHz wireless communication systems. It starts with an overview of conventional RF analog front-end architectures and their applicability to 60 GHz systems. RF nonlinearities possibly given by these architectures are also presented, with emphasis on power amplifier nonlinearities. A brief review of power amplifier models is given, and their effect on system performance presented. Then, phase noises arising from a local oscillator are investigated, with primary emphasis on their modeling procedures. This chapter ends with a brief introduction to other RF nonlinearities that may also affect system performance.

Chapter 4 discusses antenna array beamforming as a technology enabling Gbps throughput over general 60 GHz non-line-of-sight (NLOS) channels. The 60 GHz channel is briefly analyzed, and transmit/receive beamforming is shown to be a necessary technique for the 60 GHz channel. For transmit/receive beamforming, an antenna training/tracking algorithm is crucial such that the NLOS blocking issue can be solved. Two different antenna training/tracking methods are presented. One is the iterative antenna training and tracking method for adaptive antenna arrays, and the other is the divide-and-conquer training and tracking method for switched antenna arrays.

Chapter 5 discusses baseband modulation in achieving Gbps throughput at 60 GHz. We focus particularly on orthogonal frequency division multiplexing (OFDM) and single carrier (SC) block transmission (SCBT) as two major candidates in enabling high spectral efficiency transmissions. In the first part, a brief introduction to OFDM communications is given, followed by general OFDM design considerations. The challenges of designing OFDM systems for 60 GHz systems are also emphasized. The first part then uses IEEE 802.15.3c audio video (AV) OFDM as a case study and discusses various issues in baseband designs, including uncompressed video communications, physical layer equal and unequal error protection schemes, bit interleaving and multiplexing schemes, and AV OFDM modulation. The second part is devoted to SCBT with frequency-domain equalization (SC-FDE), which provides very low to very high bit rates with excellent robustness. The second part starts with a rationale for using a SC at 60 GHz and then describes how this is specified in the IEEE802.15.3c standard. The chapter continues with system aspects such as transceiver design, effect of non-idealities and equalizer design. Then, a large section is devoted to describe the signal processing functions of the SC receiver, covering acquisition, joint estimation of channel, fine carrier frequency offset and I/Q imbalance parameters, equalization, tracking and decoding.

Chapter 6 presents the current solutions, techniques and tradeoffs involved in the implementation of a high data rate 60 GHz radio in silicon from the RF front-end to the mixed-signal (analog–digital) interface with a digital baseband integrated circuit. The discussion starts with an overview of the different silicon technologies available for the implementation of 60 GHz systems, analyzing their limitations and capabilities. Given that the link margin of a wireless system is strongly dependent on the receiver's noise figure and the transmitter's $P_{1\,dB}$, the performance of currently existing 60 GHz low-noise amplifier and power amplifier solutions is reviewed in detail. Radio architectures for single- and multiple-antenna (phased array) systems are presented. Radio architectures are reviewed with emphasis on their feasibility and limitations for an integrated implementation. The current state of the art in high-speed digital-to-analog and analog-to-digital converters and modulators as important system components is analyzed in the context of their application to Gbps systems. The tradeoffs involved in a radio design for SC and OFDM modulations are discussed and implementation guidelines are provided. Finally, an outlook of the remaining challenges for the implementation of commercial 60 GHz radios is presented.

Chapter 7 covers SC hardware implementation. It starts with early-phase SC implementation examples with digital baseband, one with non-coherent detection and the other with differentially coherent detection. Then, we discuss how to implement more advanced SC systems that can comply with a certain standard, such as IEEE 802.15.3c. Readers will quickly realize that an appropriate 60 GHz system demonstration requires more than just implementation work for a given standard, and algorithm-level research, in particular the receiver side, is playing a critical role in achieving robust end-to-end 60 GHz systems.

Chapter 8 presents design consideration and implementation issues for 60 GHz OFDM hardware demonstrators. After introducing the designed OFDM physical layer and frame architecture, we present baseband processor architectures and their implementation details for both OFDM transmitters and receivers. 60 GHz wireless link demonstrations with the developed OFDM demonstrator are also highlighted. Finally, the next-generation OFDM demonstrator we have designed for wireless LAN applications and its performance evaluation are briefly introduced.

Chapter 9 discusses MAC layer design for 60 GHz communications systems. The MAC layer plays a critical role in moderating access right to the shared wireless channel. In 60 GHz wireless networks, issues related to carrier sensing, deafness and device discovery which form the major medium access control challenges in the presence of directional transmission are first discussed. Then a number of techniques to improve the MAC layer performance such as a large packet size of the order of hundreds of kilobytes, data aggregation, block-ACK and automatic repeat request (ARQ), are presented. Then the chapter delves into 60 GHz MAC design considerations to support short-range uncompressed video streaming. Finally, a performance study is presented.

Chapter 10 presents further challenges and future direction for 60 GHz communication systems.

List of Contributors

André Bourdoux
Principal Scientist
Wireless Research, SSET, IMEC
Belgium

Chang-Soon Choi
Ph.D., IEEE Member,
NTT DoCoMo communications
Laboratories Europe GmbH
Munich, Germany

Marcus Ehrig
Dipl.-Ing.
IHP microelectronics GmbH,
Frankfurt (oder), Germany

Eckhard Grass
Dr.-Ing.
IHP Microelectronics GmbH
Frankfurt (oder), Germany

Yasunao Katayama
Ph.D., Senior Technical Staff Member
IBM Research – Tokyo
Yamato Kanagawa, Japan

Maxim Piz
Dr.-Ing.
IHP microelectronics GmbH
Frankfurt (oder), Germany

Harkirat Singh
Ph.D., IEEE Member
Staff Engineer
Wireless Connectivity, Samsung
Electronics
San Jose, CA, USA

Alberto Valdes-Garcia
Ph.D., Communication and
Computation Subsystems, IBM
Research
Yorktown Heights, NY, United States

Pengfei Xia
Ph.D., IEEE Senior Member
Broadcom Corp.
San Diego, CA, USA

Su-Khiong (SK) Yong
Ph.D., Marvell Semiconductor Inc.
Santa Clara, CA, USA

1

Introduction to 60 GHz[1]

Su-Khiong (SK) Yong

1.1 What is 60 GHz?

Since the first wireless transatlantic radio wave transmission demonstration by Marconi from Cornwall, England, to Newfoundland, Canada, in 1901 (based on long wave), wireless communications have undergone tremendous growth. They were first used mainly by military and shipping companies and later quickly expanded into commercial use such as commercial broadcasting services (such as shortwave, AM and FM radio, terrestrial TV), cellular telephony, and global positioning service (GPS), wireless local area network (WLAN), and wireless personal area network (WPAN) technologies. Today, these wireless communications systems have become an integral part of daily life and continue to evolve in providing better quality and user experience. One of the recent emerging wireless technologies is millimeter-wave (mm-wave) technology. It is important to note that mm-wave technology has been known for many decades, but has mainly been deployed for military applications. Over the past 5–6 years, advances in process technologies and low cost integration solutions have made mm-wave a technology to watch and begun to attract a great deal of interest from academia, industry and standardization bodies. In very broad terms, mm-wave technology is concerned with that part of the electromagnetic spectrum between 30 and 300 GHz, corresponding to wavelengths from 10 mm to 1 mm [1], as shown in Figure 1.1. In this book, however, we will focus

---

[1]This work was done when the author was affiliated with Samsung Electronics.

60 GHz Technology for Gbps WLAN and WPAN: From Theory to Practice
Edited by Su-Khiong (SK) Yong, Pengfei Xia and Alberto Valdes Garcia
© 2011 John Wiley & Sons, Ltd

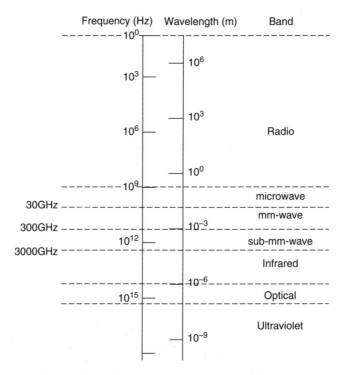

Figure 1.1   Electromagnetic spectrum allocation.

specifically on 60 GHz radio[2] which enables many new applications that are difficult if not impossible to offer by wireless systems at lower frequencies, as discussed in Section 1.3.

1.2   Comparison with other Unlicensed Systems

60 GHz technology offers various advantages over current or existing communications systems [2]. One major reason for the recent interest in 60 GHz technology is the huge unlicensed bandwidth. As shown in Figure 1.2, at least 5 GHz of continuous bandwidth is available in many countries worldwide. While this is comparable to the unlicensed bandwidth allocated for ultra-wideband (UWB) purposes [3], the 60 GHz bandwidth is continuous and less restricted in terms of power limits. This is due to the fact that UWB system is an overlay system and thus subject to very strict

---

[2]Unless otherwise specified, the terms 60 GHz and millimeter-wave are used interchangeably in this book.

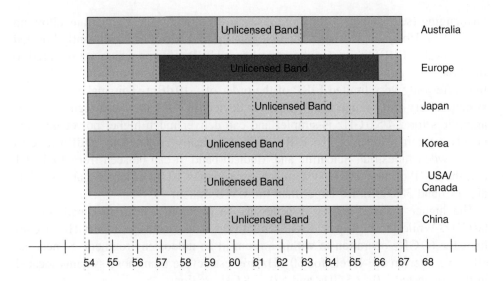

Figure 1.2   Worldwide frequency allocation for 60 GHz band and operation.

Table 1.1   Comparison of the typical implementation of 60 GHz, UWB and 802.11n systems in terms of their output power, antenna gain and EIRP output

Technology	Frequency (GHz)	PA output (dBm)	Antenna gain (dBi)	EIRP output (dBm)
60 GHz	57.0–66.0	10.0	25.0	35.0
UWB	3.1–10.6	−11.5	1.5	−10.0
IEEE 802.11n	2.4/5.0	22.0	3.0	25.0

and different regulations [4]. The large bandwidth at 60 GHz is one of the largest unlicensed bandwidths ever to be allocated. This huge bandwidth represents great potential in terms of capacity and flexibility, making 60 GHz technology particularly attractive for gigabit wireless applications (see Section 1.3).

Furthermore, 60 GHz regulation allows much higher transmit power – equivalent isotropic radiated power (EIRP) – compared to other existing WLAN and WPAN systems. Table 1.1 shows examples of typical 60 GHz, UWB and IEEE 802.11 systems that operate near the US Federal Communications Commission (FCC) regulatory limit.

The output power of a power amplifier for 60 GHz is typically limited to 10 dBm because the implementation of efficient power amplifiers at this frequency is very challenging (see Chapter 3 for more discussion) though FCC regulations allow up to 27 dBm. However, the huge antenna gain up to 40 dBi has significantly boosted the allowable EIRP limits. On the other hand, UWB systems which are required to meet the strict power spectrum mask of -41.3 dBm/MHz based on FCC regulations, thus offer only very limited EIRP of the order of -10 dBm. This makes the UWB system a very short-range and low-power device. In contrast, the design of power amplifiers for 2.4/5.0 GHz is simpler and can deliver much higher power than the 60 GHz system. However, the EIRP limit is typically confined to 30 dBm due to the crowded Industrial, Scientific and Medical band. It can be seen from Table 1.1 that the EIRP of the 60 GHz system is approximately 10 times larger than the IEEE 802.11n and 30 000 times larger than the UWB system.

The higher transmit power is necessary to overcome the higher path loss at 60 GHz. While the high path loss seems to be a disadvantage at 60 GHz, it confines the 60 GHz operation to within a room in an indoor environment. Hence, the effective interference levels for 60 GHz are less severe than those systems located in the congested 2.0–2.5 GHz and 5.0–5.8 GHz regions.

The huge bandwidth available for 60 GHz and UWB systems also simplifies the system design of these technologies. A system with much lower spectral efficiency can be designed to deliver a Gbps transmission to provide low cost and simple implementation. Table 1.2 shows the spectral efficiency required by the 60 GHz, UWB and IEEE 802.11 systems to achieve 1 Gbps transmission as well as spectral efficiency of the actual deployment of such systems. A typical 60 GHz system requires only 0.4 bps/Hz to achieve 1 Gbps, making it an ideal candidate to support very high data rate applications using simple modulation. Though the UWB system only requires 2 bps/Hz to achieve 1 Gbps, its actual deployment is limited to 400 Mbps at 1 m operating range. IEEE 802.11n-alike systems will require 25 bps/Hz in order to achieve 1 Gbps, making the extension of such system to beyond 1 Gbps unappealing in terms of cost and implementation. A more detailed discussion on the modulation choice for 60 GHz is presented in Chapter 4.

Table 1.2   Spectral efficiency comparison between 60 GHz, UWB and IEEE 802.11n technology

Technology	Bandwidth (MHz)	Efficiency@ 1 Gbps (bps/Hz)	Target data rate (Mbps)	Efficiency required (bps/Hz)
60 GHz	2000	0.5	4000	2.0
UWB	528	2.0	480	1.0
IEEE 802.11n	40	25.0	600	15.0

In addition, the huge path loss at 60 GHz enables higher-frequency reuse in each indoor environment, thus allowing a very high-throughput network. The compact size of 60 GHz radio also permits multiple-antenna solutions at the user terminal that are otherwise difficult if not impossible at lower frequencies. Compared to 5 GHz systems, the form factor of 60 GHz systems is approximately 140 times smaller and thus can be conveniently integrated into consumer electronic products.

Despite of the various advantages offered, 60 GHz based communications suffer a number of critical problems that must be solved. Figure 1.3 shows the data rates and range requirements for a number of WLAN and WPAN systems.

Since there is a need to distinguish between different standards for broader market exploitation, the 60 GHz related standards are positioned to provide gigabit rates and longer operating range than the UWB systems but shorter than that of IEEE 802.11n systems. Typically, 60 GHz systems are designed to provide multi-gigabit data rates with operating range below 20 m to support various applications as described in Section 1.3. At such rate and range, it will be a non-trivial task for 60 GHz systems to provide sufficient power margin to ensure a reliable communication link. Furthermore, the delay spread of the channel under study is another limiting factor for high-speed transmissions. Large delay spread values can easily increase the complexity of the system beyond the practical limit for equalization (see Chapter 2 for more discussion on the impact of channel on the 60 GHz design as well as choice of modulation in Chapter 5).

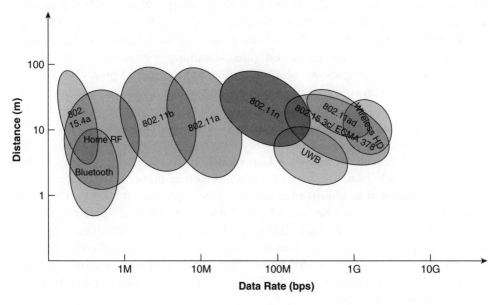

Figure 1.3 Rate versus range for WLAN and WPAN standards.

1.3   Potential Applications

With the allocated bandwidth of 7 GHz in most countries, 60 GHz radio has become the technology enabler for many gigabit transmission applications that are technically constrained at lower frequency. A number of indoor applications are envisioned, such as:

(i)   cable replacement or uncompressed high definition (HD) video streaming that enables users to wirelessly display content to a remote screen with wired equivalent quality/experience;
(ii) 'synch and go' file transfer that enables gigabytes of file transfer in a few seconds;
(iii) wireless docking stations that allow multiple peripherals (including an external monitor) to be connected without the need for frequent plugging and unplugging;
(iv) wireless gigabit Ethernet that permits bidirectional multi-gigabit Ethernet traffic;
(v)   wireless gaming that ensures high-quality performance and low latency for exceptional user experience.

All these applications have been discussed in various standards and industry alliances [5–9]. Uncompressed video streaming is emerging as one of the most attractive applications and related products based on the WirelessHD specification are currently available on the market from companies such as Panasonic, LG Electronics and Toshiba [5, 10]. In the following, we will briefly describe the technical requirements for proper uncompressed video streaming operation.

Depending on the progressive scan resolution and number of pixels per line, the data rate required varies from several hundred Mbps to a few Gbps. The latest commercially available high-definition television (HDTV) resolution is 1920 × 1080, with a refresh rate of 60 Hz. Considering RGB video formats with 8 bits per channel per pixel, the required data rate turns out to be approximately 3 Gbps, which is currently supported by the HDMI 1.1 specification. In the future, higher numbers of bits per channel (10 and 12 bits per color) as well as higher refresh rates (90 Hz, 120 Hz) are expected to improve the quality of next-generation HDTV. This easily scales the data rate to well beyond 5 Gbps. Table 1.3 summarizes data rate requirements for some current and future HDTV specifications. Furthermore, uncompressed HD streaming is an asymmetric transmission with significantly different data flow in both uplink and downlink directions. This application also requires very low latency of tens of microseconds and very low bit error probability (down to 10^{-12}) to ensure high-quality video.

Table 1.3   Data rate requirements for different resolutions, frame rates, and numbers of bits per channel per pixel for HDTV standard

Pixel per line	Active line	Frame rate per picture	Number of bits per channel per pixel	Data rate (Gbps)
1280	720	24	24	0.53
1280	720	30	24	0.66
1440	480	60	24	1.00
1280	720	50	24	1.11
1280	720	60	24	1.33
1920	1080	50	24	2.49
1920	1080	60	24	2.99
1920	1080	60	30	3.73
1920	1080	60	36	4.48
1920	1080	60	42	5.23
1920	1080	90	24	4.48
1920	1080	90	30	5.60

1.4   Worldwide Regulation and Frequency Allocation

This section discusses the current status of worldwide regulation and standardization efforts for the 60 GHz band. The regulatory bodies in the United States, Japan, Canada and Australia have already set frequency bands and regulations for 60 GHz operation, while in Korea and Europe intense efforts are currently under way. A summary of the issued and proposed frequency allocations and main specifications for radio regulation in a number of countries is given in Table 1.4. It is important to note that even though a maximum transmit power of 27 dBm is allowed in the USA, the actual transmit power may be limited by the capability of power amplifiers (PAs), especially in the case of single antennas. Typically, the maximum output of the 60 GHz PA is limited to around 10 dBm.

1.4.1   North America

In 2001, the FCC allocated 7 GHz in the 54–66 GHz band for unlicensed use [11]. In terms of the power limits, FCC rules allow emission with average power density of 9 µW/cm^2 at 3 meters and maximum power density of 18 µW/cm^2 at 3 meters, from the radiating source. These values translate to average EIRP and maximum EIRP of 40 dBm and 43 dBm, respectively. The FCC also specified the total maximum transmit power of 500 mW for an emission bandwidth greater than 100 MHz.

Table 1.4 Frequency band plan and limits on transmit power, EIRP, and antenna gain for various countries

Region	Unlicensed bandwidth (GHz)	Transmit power	EIRP (dBm)	Maximum antenna gain (dBi)
USA/Canada	7.0	500 mW or 27 dBm (max)*	40.0 (ave)+ 43.0 (max)#	33.0 (max) when 10.0 dBm TX power is used
Japan	7.0†	10 mW or 10 dBm (max)	58.0 (max)	47.0
Korea	7.0	10 mW or 10 dBm (max)	27.0 (max)	17.0‡
Australia	3.5	10 mW or 10 dBm (max)	51.7 (max)	41.8
Europe°	9.0	20 mW	57.0 (max)	30.0

Note:
*For bandwidth >100 MHz.
+Translate from average power density of $9\,\mu W/cm^2$ at 3 m.
#Translate from average power density of $18\,\mu W/cm^2$ at 3 m.
†Maximum bandwidth allowed is 2.5 GHz.
‡All devices shall transmit the Tx ID code per se. Statement is needed in user manual for antenna gain 17 dBi.
°Recommendation by the European Telecommunications Standards Institute; minimum bandwidth is 500 MHz.

Devices must also comply with the radio frequency (RF) radiation exposure requirements specified in §1.307(b), §2.1091 and §2.1093 of [11]. After taking RF safety issues into account, the maximum transmit power is limited to 10 dBm. Furthermore, each transmitter must transmit transmitter identification at least once, within 1 second interval of the signal transmission. It is important to note that 60 GHz regulations in Canada, enforced by Industry Canada Spectrum Management and Telecommunications (IC-SMT) [12], are harmonized with those of the USA.

1.4.2 Japan

In 2000, the Ministry of Public Management, Home Affairs, Posts and Telecommunications (MPHPT) of Japan issued 60 GHz radio regulations for unlicensed utilization in the 59–66 GHz band [13]. The 54.25–59 GHz band is allocated for licensed use. The maximum transmit power for unlicensed use is limited to 10 dBm [14] with maximum allowable antenna gain of 47 dBi [15]. Unlike in North America,

Japanese regulations specify that the maximum transmission bandwidth must not exceed 2.5 GHz. There is no specification for RF radiation exposure and transmitter identification requirements. For further information and latest regulation updates in Japan, interested readers are referred to [16].

1.4.3 Australia

Following the released of regulations in Japan and North America, the Australian Communications and Media Authority (ACMA) took similar steps to regulate the 60 GHz band in 2005 [17]. However, only 3.5 GHz bandwidth (59.4–62.9 GHz) is allocated for unlicensed use. The maximum transmit power and maximum EIRP are limited to 10 dBm and 51.7 dBm, respectively. The data communication transmitters that operate in this frequency band are limited to land and maritime deployments. For further information and latest regulation updates in Australia, interested readers are referred to [18].

1.4.4 Korea

In June 2005, the Millimeter-wave Frequency Study Group (MFSG) was formed under the auspices of the Korean Radio Promotion Association [19]. The MFSG has recommended a 7 GHz unlicensed spectrum (57–64 GHz) without limitations on the types of application to be used. For indoor applications, the maximum transmit power is 10 dBm, the same as in Japan and Australia, while the maximum allowable antenna gain is 17 dBi. For outdoor applications, the transmitted power is limited to −20 dBm and 10 dBm for frequency bands 57–58 GHz and 58–64 GHz, respectively, while the maximum antenna gain is 47 dBi [20]. In addition, any device using frequencies in the 57–64 GHz band shall transmit the transmitter identification code *per se*, to enable other devices to fully detect and protect against malfunctions from occurring, with the exception of any fixed point-to-point system. For further information and latest regulation updates in Korea, interested readers are referred to [21].

1.4.5 Europe

The European Telecommunications Standards Institute (ETSI) and European Conference of Postal and Telecommunications Administrations (CEPT) have been working closely to establish a legal framework for the deployment of unlicensed 60 GHz devices. In general, the 59–66 GHz band has been allocated for mobile services without specific decisions having been made as to the regulations. The CEPT Recommendation T/R 22-03 provisionally allocated (and later withdrew) the 54.25–66 GHz band for terrestrial and fixed mobile systems [22].

In 2004, the European Radiocommunications Committee (ERC) considered the used of the 57–59 GHz band for fixed services without requiring frequency planning [23]. Later, the Electronic Communications Committee (ECC) within CEPT recommended the use of point-to-point fixed services in the 64–66 GHz band [24]. Later, ETSI proposed 60 GHz regulations to be considered by the ECC for WPAN applications [25]. Under this proposal, 9 GHz of unlicensed spectrum is allocated for 60 GHz operation. This band represents the union of the bands currently approved and proposed among the major countries as shown in Figure 1.2. In addition, a minimum spectrum of 500 MHz is required for the transmitted signal with maximum EIRP of 57 dBm. No specification is given for the maximum transmit power and maximum antenna gain.

In October 2009, the CEPT recommended a maximum EIRP of 25 dBm with a maximum spectral power density of -2 dBm/MHz for outdoor applications, though a fixed outdoor installation is not allowed [26]. For indoor applications a maximum EIRP of 40 dBm with a maximum spectral power density of 13 dBm/MHz is specified [26]. It is unclear when the final regulation will be in place, but the current trend seems encouraging for the deployment of 60 GHz technology. For further information and latest regulation updates in the Europe, interested readers are referred to [27].

1.5 Industry Standardization Effort

The first international industry standard that covers the 60 GHz band is the IEEE 802.16 standard for local and metropolitan area networks [28]. However, this is a licensed band and is used for line-of-sight (LOS) outdoor communications for last mile connectivity. In Japan, two standards related to the 60 GHz band were issued by Association of Radio Industries and Business (ARIB): the ARIB-STD T69 [29] and ARIB-STD T74 [30]. The former is the standard for mm-wave video transmission equipment for specified low-power radio stations (point-to-point systems), while the latter is the standard for mm-wave ultra high-speed WLANs for specified low power radio stations (point-to-multipoint systems). Both standards cover the 59–66 GHz band defined in Japan.

Interest in 60 GHz radio continued to grow with the formation of multiple international mm-wave standards groups and industry alliances. In March 2005, the IEEE 802.15.3c Task Group (TG3c) was formed to develop a mm-wave based alternative physical layer (PHY) for the existing IEEE 802.15.3 WPAN standard 802.15.3-2003 [31]. In August 2006, ECMA TC-48 (formerly known as TC32-TG20) began an effort to standardize medium access control (MAC) and PHY for high-speed, short-range communications using the 60 GHz unlicensed frequency band for bulk data applications and for multimedia streaming applications [7]. In October 2006, the formation of the WirelessHD consortium was announced with a number of key consumer electronics companies to deliver a specification for high-speed, high-quality uncompressed audio/video (A/V) streaming using 60 GHz technology [5]. In the latest development, the Wireless Gigabit Alliance (WiGig) was formed in May 2009 to establish a unified specification for 60 GHz wireless technology in order to create a truly global ecosystem of interoperable products for a diverse range of applications [9]. In this section we briefly describe a number of standardization efforts.

1.5.1   IEEE 802.15.3c

The alternative PHY of IEEE 802.15.3c is aimed at supporting a minimum data rate of 2 Gbps over a few meters with optional data rates in excess of 3 Gbps. This is the first standard that addresses multi-gigabit short-range wireless systems. The IEEE 802.15.3c standard was ratified in September 2009 [31]. Three PHYs are defined in the specification, namely single carrier (SC) PHY, high speed interface (HSI) orthogonal frequency division multiplexing (OFDM) PHY and audio video (AV) OFDM PHY. The need for three PHYs is due to the inherent advantages of each PHY in supporting specific applications. SC PHY is designed to support low-cost and low-power mobile devices; HSI PHY is used for low-latency, high-speed bidirectional data transmission, while AV PHY is optimized for AV specific applications. The key PHY features of IEEE 802.15.3c are summarized in Table 1.5.

In order to promote coexistence among these PHY modes, common mode signaling (CMS) is defined which is an SC-based $\pi/2$ binary phase shift key (BPSK)

Table 1.5   Summary of the PHY modes in the IEEE 802.15.3c standard

Feature	SC-FDE	HSI OFDM	AV OFDM
Constellation	BPSK, (G)MSK, QPSK, 8PSK, 16QAM	QPSK, 16 QAM 64QAM	QPSK, 16QAM
Data rate	25.3 Mbps–5.1 Gbps	31.5 Mbps–5.67 Gbps	0.95–3.8 Gbps
Coding	Reed Solomon, LDPC	LDPC	Reed Solomon and Convolutional Coding
UEP Support	Yes	Yes	Yes
Training sequence	Golay Code	Golay Code	M-Sequence Barker-13 chip
Beamforming	Yes	Yes	Yes
Occupied bandwidth	1.782 GHz	1.782 GHz	1.76 GHz (HRP) 92 MHz (LRP)

with low data rate (25 Mbps). CMS is used in piconet coordinator capable devices to send/receive a CMS synch frame in order to avoid interference between two or more operating piconets. Beamforming is supported by all the three PHY modes for heterogeneous antenna types. The beamfoming employs a two-level mechanism to find the optimum transmit and received beams that enable high data rate transmission. In addition, 15.3c specifies an unequal error protection (UEP) support for uncompressed video transmission. UEP can be achieved in both PHY and MAC layers. UEP at PHY protects the most significant bit (MSB) and least significant bit (LSB) in a subframe unequally by applying different coding and/or constellation mapping. On the other hand, UEP at MAC protects an aggregated frame which consists of MSB subframes, LSB subframes, or both MSB and LSB subframes, by using either different forward error corrections or different modulation and coding schemes. We will revisit the UEP PHY and MAC in more detail in Chapters 5 and 9, respectively.

1.5.2   ECMA 387

ECMA TC48 developed the ECMA 387 specification for high-rate 60 GHz PHY, MAC and HDMI PAL for short-range unlicensed communications. The first edition of the ECMA 387 was ratified in December 2008 and subsequently submitted to ISO/IEC JTC 1's fast-track procedure to turn ECMA-387 into ISO/IEC 13156 by the end of 2009 [32]. Three types of devices (i.e. Type A, Type B and Type C) are specified in ECMA 387 based on the complexity and power consumption. Type A represents the most complex and power-hungry device types, intended to deliver video/data even without LOS by employing beamforming. Type B represents devices of moderate complexity and power consumption and is designed to deliver video/data in LOS without using beamforming. Finally, Type C devices are the least complex and have lowest power consumption, and are used for data delivery over very short range (less than 1 meter). In addition, three mandatory PHYs are also defined in the specification, namely SC block transmission (SCBT), SC with differential binary phase shift keying (DBPSK) and SC with binary amplitude shift keying (BASK, also known as OOK). These mandatory PHYs are mapped into Type A, B and C devices as shown in Table 1.6.

In order to promote coexistence among these device types, during device discovery, which takes place in the discovery channel, Type A devices need to support SCBT, DBPSK and OOK modes while Type B devices need to support both DBPSK and OOK. While ECMA 378 specifies interoperability and coexistence among these three device types, the overall protocol and implementation complexity have dramatically increased, especially for Type A devices, as compared to IEEE 802.15.3c. Beamforming is mandatory for Type A devices while Type B devices may support beamforming training of Type A devices by feeding back the best adaptive weight

Table 1.6 Summary of device types and associated PHY modes in ECMA 378

	Type A	Type B	Type C
Mandatory modes	SCBT, DBPSK, OOK	DBPSK, OOK	OOK
Optional modes	25.3 Mbps–5.1 Gbps OFDM, DQPSK, 4ASK	31.5 Mbps–5.67 Gbps DQPSK, Dual-AMI 4ASK	0.95–3.8 Gbps 4ASK
Coding	Reed Solomon and Convolutional Coding	Reed Solomon	Reed Solomon
UEP support	Yes	Yes	No
Beamforming	Yes	No[+]	No
Beacon transmission	SCBT	DBPSK, SCBT*	N/A
DRP transmission	SCBT, OFDM, DBPSK, OOK, 4ASK	DBPSK, DQPSK, Dual AMI, OOK, 4ASK	OOK/4ASK

Note:
**SCBT is used for transmit only
[+]Assist in feeding back the transmit beamforming information and/or sending antenna training sequences for receive beamforming of Type A devices.

vector (AWV) for transmit beamforming and/or sending antenna training sequences for receive Beamforming. Similar to IEEE 802.15.3c, UEP is also supported at both PHY and MAC layers.

1.5.3 WirelessHD

The WirelessHD consortium developed the WirelessHD 1.0 specification released in January 2008 [5]. An overview of the WirelessHD specification 1.0 can be found in [33]. Unlike 15.3c and ECMA-378, WirelessHD only uses OFDM modulation with a smaller number of modes. A close comparison between the WirelessHD 1.0 specification [33] and AV OFDM PHY in 15.3c [31] reveals a lot of similarities. Both consist of OFDM based high-rate PHY (HRP) and low-rate PHY (LRP) that share the same frequency band plan, baseband and general parameters. Two types of beamforming are specified, namely, explicit beamforming and implicit beamforming. Explicit beamforming is mandatory and involves a beam search phase, which is used to estimate the AWV at both the transmitter and receiver, and a beam tracking phase, which is used to track the changes of the existing AWVs over time as a result of slow variation of the channel. On the other hand, implicit beamforming employs a beambook approach whereby a set of common beam vectors are maintained at both the transmitter and receiver. WirelessHD 1.0 also supports UEP for uncompressed video streaming applications. WirelessHD has gained some support from the consumer electronics industry. Companies such as LG Electronics, Panasonic, Toshiba and Gefen have already brought WirelessHD products to the market [10, 34], which represents a major milestone in commercializing 60 GHz technology. Recently, WirelessHD announced a next-generation WirelessHD specification that can support data rates up to 28 Gbps for 3D TV and 4K resolution support [35]. In addition, the next generation will also support data for low-power portable devices at a data rate of 1 Gbps.

1.5.4   IEEE 802.11.ad

IEEE 802.11ad was formed in January 2009 as an amendment to the existing IEEE 802.11-2007. This amendment defines standardized modifications to both the 802.11 PHY and the 802.11 MAC to enable very high-throughput operation in the 60 GHz frequency band [36]. This amendment specifies a maximum MAC service access point throughput of at least 1 Gbps while maintaining the network architecture of the 802.11 system as well as backward compatibility with the 802.11 management plane. In addition, IEEE 802.11ad is intended to define a mechanism for fast session transfer between 2.4/5 GHz and 60 GHz operation bands, while coexisting with other systems in the 60 GHz band such as IEEE 802.15.3c and ECMA-378. IEEE 802.11ad is expected to be completed by 2012.

1.5.5   Wireless Gigabit Alliance

The Wireless Gigabit Alliance (WiGig) was formed in May 2009 with broad support from the personal computer, consumer electronics, semiconductor and mobile handheld industries. The key objective of WGA is to establish a unified specification for 60 GHz wireless technologies that can support diverse applications as specified in Section 1.3. The WGA specification was released in May 2010.

1.6   Summary

The combination of high EIRP limit, huge bandwidth, and harmonized regulation and frequency allocation globally has positioned 60 GHz in the forefront of Gbps wireless communications. This can be demonstrated by the immense standardization effort and industry alliance formation to promote 60 GHz technology. Despite the tremendous progress made in 60 GHz technology in the past decade, the challenges of full-scale commercialization still remain, particularly in providing low-cost, low-power and robust 60 GHz products. It is the aim of this book to highlight and address some of the key issues related to 60 GHz technology from propagation, to PHY and MAC design as well as integrated circuit implementation of 60 GHz.

References

[1] Oliver, A.D. (1989) Millimeter wave systems – past, present and future. *IEE Proceedings*, 136(1), 35–52.

[2] Yong, S.K. and Chong, C.C. (2007) An overview of multigigabit wireless through millimeter wave technology: Potentials and technical challenges. *EURASIP J. Wireless Commun. and Networking*, article ID 78907, 10 pp.

[3] FCC First Report And Order (2002) http://hraunfoss.fcc.gov/edocs_public/attachmatch/FCC-02-48A1.pdf

[4] Chong, C.C., Watanabe, F. and Inamura, H. (2006) Potential of UWB technology for the next generation wireless communications. *Proceedings of IEEE International Symposium on Spread Spectrum Techniques and Applications (ISSSTA'06)*, pp. 422–429, Manaus, Brazil, August.

[5] Wireless High-Definition (WirelessHD) (2008) http://www.wirelesshd.org/

[6] IEEE 802.15 (2005) WPAN Millimeter Wave Alternative PHY Task Group 3c (TG3c) http://www.ieee802.org/15/pub/TG3c.html

[7] ECMA TC48 (2006) High rate short range wireless communication. http://www.ecma-international.org/memento/TC48-M.htm

[8] IEEE 802.11 Very High Throughput (VHT) Study Group (2008) http://www.ieee802.org/11/Reports/vht update.htm

[9] Wireless Gigabit Alliance (WiGig) (2009) http://wirelessgigabitalliance.org/news/wigig-alliance-publishes-multi-gigabit-wireless-specification-and-launches-adopter-program/

[10] WirelessHD (2009) The first 60 GHz standard now available in consumer electronics products worldwide. http://www.wirelesshd.org/pdfs/WiHD%20CEDIA%20SEPT0901009_FINAL.pdf

[11] FCC Code of Federal Regulation (2004) *Code of Federal Regulation title 47 Telecommunication, Chapter 1, part 15.255*, October.

[12] Spectrum Management and Telecommunications (2005) *Radio Standard Specification-210, Issue 6, Low-power Licensed-exempt Radio Communication Devices (All Frequency Bands): Category 1 Equipment*, September.

[13] Regulations for Enforcement of the Radio Law 6-4-2 Specified Low Power Radio Station (12) 59–66 GHz band.

[14] MPT (1989) Specified low power radio station 10. Millimeter wave transmission. *MPT Bulletin*, mo. 42/1989.

[15] Ordinance for Regulating Radio Equipment 49-14-6 Specified Low Power Radio Station.

[16] Ministry of Internal Affairs and Communications. http://www.soumu.go.jp/english/index.html

[17] Australian Communications and Media Authority (ACMA) (2005) *Radiocommunications (Low Interference Potential Devices) Class License Variation 2005 (No. 1)*, August.

[18] Australian Communications and Media Authority (ACMA) http://www.acma.gov.au

[19] Korean Frequency Policy & Technology Workshop (2005), Session 7, pp. 13–32, November.

[20] Ministry of Information and Communication of Korea (2006) *Frequency Allocation Comment of 60 GHz Band*, April.

[21] Korean Communications Commission http://www.kcc.go.kr

[22] CEPT Recommendation T/R 22-03 (1990) *Provisional Recommended Use of the Frequency Range 54.25-66 GHz by Terrestrial Fixed and Mobile Systems*. Athens, pp. 1–3, January. http://www.ero.dk/documentation/docs/doc98/official/pdf/TR2203E.PDF

[23] ERC Recommendation 12-09 (2004) *Radio Frequency Channel Arrangement for Fixed Service Systems Operating in the Band 57.0–59.0 GHz Which Do Not Require Frequency Planning, The Hague 1998 revised Stockholm*, October.

[24] ECC Recommendation 05-02 (2005) *Use of the 64–66 GHz Frequency Band for Fixed Services*, June.

[25] ETSI DTR/ERM-RM-049 (2006) *Electromagnetic Compatibility and Radio Spectrum Matters (ERM): System Reference Document: Technical Characteristics of Multiple Gigabit Wireless Systems in the 60 GHz Range*, March.

[26] ERC Recommandations 70-03 (1997) http://www.erodocdb.dk/Docs/doc98/official/pdf /REC7003E.PDF

[27] European Radiocommunications Office http://www.ero.dk

[28] IEEE Standard 802.16 (2001) *IEEE Standard for Local and Metropolitan Area Networks – Part 16 – Air Interface for Fixed Broadband Wireless Access Systems*.

[29] ARIB STD-T69 (2005) *Millimeter-Wave Video Transmission Equipment for Specified Low Power Radio Station*.

[30] ARIB STD-T74 (2005) *Millimeter-Wave Data Transmission Equipment for Specified Low Power Radio Station (Ultra High Speed Wireless LAN System*, November.

[31] IEEE P802-15-3c-D13 (2009) *IEEE P802-15-3c-D13 Part 15.3: Wireless Medium Access Control (MAC) and Physical Layer (PHY) Specifications for High Rate Wireless Personal Area Networks (WPANs): Amendment 2: Millimeter-wave based Alternative Physical Layer Extension*.

[32] ECMA-387 (2009) *High Rate 60 GHz PHY, MAC and HDMI PAL*. http://www.ecma-international.org/activities/Communications/tc48-2009-005.doc

[33] WirelessHD (2009) *An Overview of WirelessHD Specification 1.0*. http://www.wirelesshd.org /pdfs/WirelessHD_Full_Overview_071009.pdf

[34] EE Times (2009) 60 GHz gains traction at CES. http://www.eetimes.com/news/latest /showArticle.jhtml?articleID=212800003

[35] WirelessHD (2010) WirelessHD Next Gen supports 3DTV, HDCP 2.0, data applications and data rates in excess of 10 Gbps. http://www.wirelesshd.org/pdfs/WiHD%20Next%20Gen %20Jan10%20FINAL.pdf

[36] IEEE 802.11ad PAR (2009) https://development.standards.ieee.org/get-file/P802.11ad.pdf?t =29195900024

2

60 GHz Channel Characterizations and Modeling[1,2]

Su-Khiong (SK) Yong

2.1 Introduction to Wireless Channel Modeling

A reliable wireless system design and system performance evaluation requires a realistic channel model, which closely resembles real propagation environments. Accurate propagation channel modeling is the first major step to be carried out before any system design, as evident as part of the many new technologies and wireless standards development such as IEEE 802.11n [1], IEEE 802.15.3c [2], IEEE 802.11ad [3] and ITU-R IMT Advanced [4].

A good radio channel model can provide detailed insight into the complex radio wave propagation mechanisms as well as allowing study of the achievable performance from both theoretical and simulation standpoints. When implemented as channel simulators, the performance of different transmission technologies and signal processing algorithms can be compared without the need to perform expensive field trials or tests for every scenario under investigation. Unfortunately, the mechanisms that govern radio propagation in a wireless communication channel are very complex and diverse, which makes channel modeling interesting as a research subject. While it is important to keep the channel models as accurate

[1]Portions of this chapter are reproduced with permission from S.K. Yong, "TG3c Channel Modelling Sub-Committee Final Report," © March 2007 and A. Maltsev, "Channel Models for 60 GHz WLAN Systems," IEEE © 2010.
[2]This work was done when the author was affiliated with Samsung Electronics.

as possible in order to sufficiently capture the key channel effects, the models developed must also be simple enough to be implemented. This is particularly important in system simulation that involves multiple-antenna systems such as multiple-input multiple-output (MIMO) in order to reduce the overall simulation time. Hence, a good channel modeling approach involves a tradeoff between accuracy and simplicity of the system under investigation.

Typically, the type of channel model that is desired depends critically on carrier frequency, bandwidth, the type of environment and system under consideration. In the context of 60 GHz, for example, indoor environments with wideband systems are usually considered. In addition, due to the directional nature of 60 GHz communications systems, the channel models developed also include the effect of the antenna under study, referred to as the *radio channel* model. Removing the effect of the antenna from the channel model results in a *propagation channel* [5], whereby the model is valid for any antenna types used in the channel model. However, this requires a double directional channel modeling approach, which is difficult to implement, especially in the 60 GHz band. To the best of our knowledge, all the 60 GHz channel models to date are radio channels and can be modeled using various approaches as described in Section 2.2

2.2 Modeling Approach and Classification of Channel Model

The requirements for modeling various types of wireless propagation channels have resulted in a large number of different modeling approaches reported in the literature [6–8]. The complex phenomenon whereby a transmitted signal propagates through the wireless channel and arrives at the receiver, typically via many different paths, is known as *multipath*. During this process, the signal undergoes various propagation mechanisms such as reflection off the wall, scattering around furniture and/or diffraction around the corners of buildings. Therefore, many different types of simplifications and approximations can be used to model the wireless communications channels, depending on their usage.

In general, propagation channel models can be broadly divided into two main categories, namely *deterministic* and *stochastic* (see Figure 2.1). These categories differ in terms of their usage and the underlying data type used in deriving them.

2.2.1 Deterministic Modeling

Deterministic model approaches can be further divided into three subgroups: the *closed-form* approach, *empirical* (based on measurement) approach and *ray tracing* approach. A good example of the closed-form approach is the two-path signal model [9], a very simple model that allows theoretical and analytical analysis for

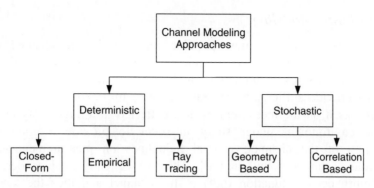

Figure 2.1 Classification of channel modeling approaches. These modeling approaches are based on tradeoff between accuracy and simplicity.

different transmission schemes. However, this approach is too conservative to represent the realistic scenario of the environment under study. On the other hand, the empirical approach extracts channel parameters from the data collected in channel measurements conducted in a specific environment under study. Hence, the empirical approach is very accurate, despite its mathematical simplicity. But such accuracy comes at the expense of high complexity due to the large amount of measurement data needed. Furthermore, the validity of the empirical model is very site-specific and difficult to relate to the physical processes of the propagation. A number of empirical models for 60 GHz has been developed and reported in the literature [10–14]. In the ray tracing approach, advanced electromagnetic theory and simulation tools (such as the uniform theory of diffraction and finite difference time domain) are used in developing the desired channel model. A good example of this approach is the use of the ray launching method. This method first defines a site-specific environment based on a detailed map. The location, size and orientation of the access point (AP), station (STA) and various objects (such as furniture and humans) are placed in the environment according to the actual setup. A more complete solution characterizing the properties of the object (such as permittivity and attenuation) can also be applied to the simulation tool in order to obtain more accurate propagation characteristics for the specified geometry setup. Then, a known signal is transmitted and the received signal computed in order to characterize the propagation mechanism. Ray tracing is highly environment-specific and its accuracy is heavily dependent on the availability of the materials and topographical database to be utilized in the modeling [15]. As such, ray tracing is usually considered as the most complex and computationally expensive of the deterministic modeling approaches. Some ray tracing models for 60 GHz systems have been reported in the literature; see [16–19] and the references therein.

2.2.2   Stochastic Modeling

Stochastic modeling is the most popular approach in the channel modeling community since it offers good a tradeoff between complexity and accuracy. Compared to deterministic modeling, stochastic modeling has lower complexity and yet can provide sufficiently accurate channel information. Stochastic models are derived based on the measurement data collected in a large array of locations within the environment under study in order to provide a good statistical representation of the channel; general guidelines on measurement techniques and procedures can be found in [20]. This subsequently enables derivation of the probability density function (pdf) of the channel and thus the key channel parameters. The pdf and its associated channel parameters will be used to generate the channel impulse response (CIR) needed for simulation purposes. One of the key features of stochastic modeling is that it can be tuned to replicate various other different scenarios such as non-line-of-sight (NLOS) or other environments that share the same pdf by setting the channel parameters to the appropriate values.

The stochastic approach can also be divided into two subgroups: the *geometry based stochastic model* (GBSM) and the *correlation based model*. In the literature, the GBSM is sometimes referred to as the ray-based model. In GBSM, scatterers are distributed in a specific manner on a certain predefined geometrical shape such as an ellipse [21] or circle [22] between the transmitter (TX) and receiver (RX) in order to imitate the effect of wave propagation. Typically, a single bounce is assumed for simplicity [23], and a double bounce can sometimes be used for more advanced modeling [24]. The channel amplitude, phase, angle and time information for each path can be derived based on the superposition of the paths that arrive at the RX as they propagate from the TX and are reflected off the scatterers.

On the other hand, the correlation-based modeling approach, which has been widely used to model multiple antenna systems, employs the channel second-order statistics. In this approach, the transfer functions of each TX and RX antenna element pair, as well as their signal correlations, are characterized. The generic full correlation model describes the spatial behavior of a MIMO channel using a full correlation matrix given by

$$\mathbf{R} = E\{\text{vec}(\mathbf{H})\text{vec}(\mathbf{H})^H\}, \tag{2.1}$$

where $\mathbf{H}$ is a channel matrix of size $N_{\text{RX}} \times N_{\text{TX}}$, where N_{RX} is the number of RX antennas and N_{TX} the number of TX antennas. The model requires a total of $(N_{\text{TX}}N_{\text{RX}})^2$ real-valued parameters, which makes it too complex when the number of antennas is large, especially in the context of 60 GHz technology. Furthermore, the relationship between the elements of the correlation matrix and the physical interpretation of the channel is difficult to establish under the full correlation model [25]. A simplified version of the correlation based model is the Kronecker based model [26], which is used in IEEE 802.11n model. The Kronecker model is computationally simpler than the full correlation based model, especially when the number of antennas is large since the directional properties of the channel at the TX and RX are assumed to be independent.

The pros and cons of the GBSM and correlation based stochastic channel models are mainly related to the implementation complexity, accuracy and whether the spatial-temporal correlation is explicitly available or implicitly inferred from the channel model. The latter can provide a good platform for theoretical framework development. For instance, the GBSM channel model is not very useful for theoretical analysis since the channel coefficients generated do not explicitly present the correlation properties. This makes it very difficult to relate the simulation results to the theoretical analysis. On the other hand, the correlation model, which only describes the second-order statistics, provides no physical intuition for the propagation. In addition, the simplicity of the correlation based approach is down to the smaller number of input parameters as compared to the GBSM model. However, such simplicity is traded off by the applicability of the model to only the average correlation behavior of the channel, which is antenna array dependent.

2.3 Channel Characterization

For all the modeling approaches discussed in Section 2.2, two important channel properties can be characterized, namely *large-scale* and *small-scale* channel characterizations, which will be discussed in Sections 2.3.1 and 2.3.2, respectively.

2.3.1 Large-Scale Channel Characterization

Large-scale channel characterization consists of *path loss* (PL) and *shadowing* effects, as illustrated in Figure 2.2.

2.3.1.1 Path Loss

The PL is defined as the ratio of the received signal power to the transmit signal power, which describes the attenuation of the mean power as a function of distance traveled. PL is very important for link budget analysis and network planning in order to ensure that the actual deployment can meet the target coverage. The PL at 60 GHz is much more severe than at lower frequencies since the free space PL at 60 GHz alone increases by approximately 22 dB compared to 5 GHz band. Furthermore, the

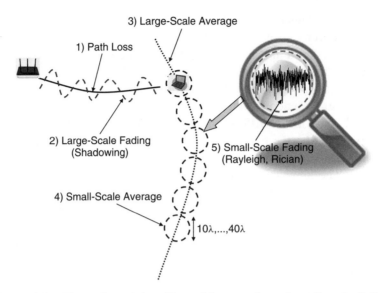

Figure 2.2   Illustration of the effect of large-scale and small-scale fading.

PL at 60 GHz is subject to additional losses due to oxygen absorption and rain attenuation. As shown in Figure 2.3, the oxygen absorption peaks at 60 GHz with 15 dB attenuation per kilometer (with 10 dB/km around the 8 GHz band centered around 60 GHz), while rain attenuation adds losses of a few decibels, depending on the rain intensity. This poses a severe challenge in delivering a gigabit wireless transmission with reliable link margin and makes 60 GHz a promising candidate for indoor rather than outdoor applications.[3]

Furthermore, the difference in PL for a wideband system such as ultra-wideband (UWB) [27, 28], and 60 GHz, as compared to the narrowband system (e.g. IEEE 802.11a/b/g/n), is that the PL at UWB and 60 GHz is both *distance* and *frequency* dependent. For example, [28] reported a frequency dependent PL model for the UWB channel. Unfortunately, 60 GHz PL modeling with frequency dependence has not yet been reported in the literature. In order to simplify the models, it is assumed that the mid-band frequency point is used and only distance dependence PL is modeled in this book (i.e. ignoring the frequency dependence PL). The PL as a function of distance, d, is thus given by

$$PL(d)[dB] = \overline{PL}(d)[dB] + X_\sigma[dB], \tag{2.2}$$

---

[3]Outdoor application for 60 GHz is mainly backhaul which employs highly directional antennas that point to each other in a clear line-of-sight (LOS) scenario.

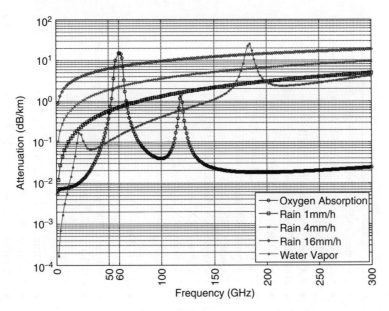

Figure 2.3 Free space PL, oxygen absorption, and rain attenuation comparison across the frequency of interest.

where $\overline{PL}(d)$ is the average PL and X_σ is the shadowing fading. In general, $\overline{PL}(d)$ can be expressed as

$$\overline{PL}(d)[dB] = \underbrace{PL(d_0)[dB]}_{\substack{PL \text{ at reference} \\ \text{distance}}} + \underbrace{10n\log_{10}\left(\frac{d}{d_0}\right)}_{\substack{PL \text{ exponent at} \\ \text{relative distance, } d}} + \sum_{q=1}^{Q} X_q, \quad \text{for } d \geq d_0,$$

(2.3)

where d_0 and n denote the reference distance and PL exponent, respectively. The term X_q accounts for the additional attenuation due to specific obstruction by objects. The PL exponent is obtained by performing least squares linear regression on the logarithmic scatter plot of averaged received power versus distance to Equations (2.2)–(2.3). Prior to that, the data is segmented into LOS and NLOS scenarios, respectively. Typically, a value of $d_0 = 1$m is used as the reference even though different values are also commonly used in many measurements.

A comprehensive list of the values of n and the standard deviation, σ_S (see also Section 2.3.1.2), reported in the literature is given in Table 2.1. Unless otherwise specified, the $PL(d_0)$ values are referred to $d_0 = 1$m.

Table 2.1 The PL exponent, n, and standard deviation for shadowing, σ_S, reported in the literature

Ref	Center freq. [GHz]	Environment	Scenario	n	σ_S	Comments
[29]	60	Hallway (10.2 × 2.1 × 4.3 m)	LOS	1.88– 2.00	8.6	• TX open-ended waveguide with 6.7 dB • Half power beamwidth (HPBW) 90° azimuth and 125° elevation • RX horn antenna with 29 dB • HPBW 7° in azimuth and 5.6° in elevation
[30]	60	Typical office laboratory	LOS and NLOS	2.10	7.9	• TX and RX horn 25 dBi • Typical office cubical and chairs
[31]	59.9	Corridor (45.0 × 2.2 m)	LOS	1.87 1.88	N/A	• TX omni, RX directional (19.5 dBi, 15°) • TX omni, RX omni
		Amphitheater (18/12 × 15 m)	LOS	0.78 1.27		• TX omni, RX directional (19.5 dBi, 15°) • TX omni, RX omni
		Grass field (two sides with buildings)	LOS	1.90		• TX omni, RX high AP, low AP
[32]	60	Laboratory (19.5 × 7.5 m)	LOS NLOS	1.80 2.00	N/A	• $d_0 = 1.5$ m
[33]	60	Empty medium size room (6 × 16 m)	LOS	2.24	5.2	• TX omni, RX horn (20 dBi, 3 dB beamwidth 20°) • $PL(d_0) = 65.7$ dB
		Cell office (13 × 12 m)	LOS	1.32	10.1	• TX omni, RX horn (20 dBi, 3 dB beamwidth 20°) • $PL(d_0) = 78.9$ dB
			NLOS	2.83	7.10	• TX omni, RX horn (20 dBi, 3 dB beamwidth 20°) • $PL(d_0) = 61.7$ dB

						Description
		Open office (14 × 30 m)	LOS	1.77	6.00	• TX omni, RX horn (20 dBi, 3 dB beamwidth 20°) • $PL(d_0) = 71.9$ dB
			NLOS	3.83	7.60	• TX omni, RX horn (20 dBi, 3 dB beamwidth 20°) • $PL(d_0) = 57.5$ dB
		Open office with partition walls (30 × 20 m)	LOS	1.16	5.40	• TX omni, RX horn (20 dBi, 3 dB beamwidth 20°) • $PL(0) = 84.6$ dB
			NLOS	3.74	8.60	• TX omni, RX horn (20 dBi, 3 dB beamwidth 20°) • $PL(d_0) = 56.1$ dB
		Corridor (40 × 2 m)	LOS	2.29	8.4	• TX horn, RX horn (20 dBi, 3 dB beamwidth 20°) • $PL(d_0) = 69.7$ dB • Vertical to vertical
				0.87	3.6	• TX horn, RX horn (20 dBi, 3 dB beamwidth 20°)) • $PL(d_0) = 115.5$ dB • Vertical to horizontal
[34]	60	Office	LOS	2.00	1.25	• TX omni, RX horn (20 dBi, 3 dB beamwidth 20°) • TX omni at 1.4 m height, RX omni at 1.4 m height • $PL(d_0) = 68.3$ dB
[35]	60	Laboratory	LOS	1.20	2.70	• TX omni at 1.9 m height, RX omni at 1.4 m height • $PL(d_0) = 83.8$ dB
				0.60	2.00	
				0.20	1.30	• TX omni at 2.4 m height, RX omni at 1.4 m height • $PL(d_0) = 87.8$ dB

(continued overleaf)

Table 2.1 (*continued*)

Ref	Center freq. [GHz]	Environment	Scenario	n	σ_S	Comments
			NLOS	5.40	3.90	• TX omni at 1.4 m height, RX omni at 1.4 m height • $PL(d_0)=34.8$ dB
				3.80	3.30	• TX omni at 1.9 m height, RX omni at 1.4 m height • $PL(d_0)=56.7$ dB
				2.70	2.70	• TX omni at 2.4 m height, RX omni at 1.4 m height • $PL(d_0)=71$ dB
			LOS	0.40	1.00	• TX fan beam (16.5 dB with 70° HPBW) at 2.5 m height, RX omni at 1.4 m height • $PL(d_0)=79.7$ dB
				2.10	0.80	• TX fan beam (16.5 dB with 70° HPBW) at 2.5 m height, RX fan beam (16.5 dB with 70° HPBW) at 1.4 m height, • $PL(d_0)=67.0$ dB
				2.00	0.60	• TX fan beam (16.5 dB with 70° HPBW) at 2.5 m height, RX pencil beam (24.4 dB with 8.3° HPBW) at 1.4 m height, • $PL(d_0)=67.4$ dB
[36]	61.7	Corridor	LOS	1.64	2.53	• TX horn (22.7 dB) at 1.25 m, RX horn (5 dB) at 1.23 m
	61.3	Hall	LOS	2.17	0.88	• TX horn (5 dB) at 1.67 m, RX horn (5 dB) at 1.66 m
			NLOS	3.01	1.55	

Ref	Freq (GHz)	Environment		Value		Notes
[37]	21.6	Corridor	LOS	1.20	N/A	• (30 m ×45 m) floor size with rooms and hallways of various sizes.
	37.2	Corridor	LOS	1.65	N/A	• Steel doors, double plaster board internal wall and 1 ft² tile floor
	21.6	1–4 Wall Obstructions	NLOS	2.95	N/A	• TX, RX biconical omni, both at 1.5 m height
	37.2	1–4 Wall Obstructions	NLOS	3.30	N/A	
[38]	94	In a hall (17 × 14.5) m, a room (12.6 × 6) m and 3 m width corridor	LOS	1.20–1.80	N/A	• Plasterboard walls and concrete floor
			NLOS	3.6–4.1	N/A	• TX horn with 25 dBi (3 dB beamwidth 10°), RX slot with 11 dBi; both at 0.9 m height
[39]	40	Open concept office (37 × 55 m)	LOS	1.50	NA	• Furnished with 1.22 m high semi-permanent partitions dividing many work spaces • TX omni, RX omni
[40]	60	Office	NLOS	4.00	NA	• TX HPBW 9.8° and 9.1° in E- and H-planes, respectively, and a gain of about 23 dB
			LOS	2.1		• RX omni biconical with gain of 5 dB
			NLOS	3.5		

The PL exponent, n, for 60 GHz based measurements ranges from 0.40 to 2.10 and from 1.97 to 5.40 for LOS and NLOS, respectively, in various different indoor environments. Specifically, in a typical office environment, the value of n for LOS and NLOS ranges from 1.16 to 2.17 and 2.83 to 4.00, respectively, while in a laboratory environment, n takes values in the ranges 0.20–2.10 and 2.00–5.10 in LOS and NLOS, respectively. Furthermore, in a corridor environment, n varies from 0.87 to 2.29 for LOS. In general, a more reflective environment such as a laboratory yields higher n values in NLOS since the reflected paths arriving at the receiver are attenuated more compared to an office scenario where diffraction usually occurs. The small values of $n(< 2)$ observed in LOS scenarios can be justified by the presence of waveguiding and reverberation effects, which cause an increase in power levels by multipath aggregation.

Several observations can be made on the basis of Table 2.1. Firstly, it can be observed that even in the same type of environment, the derived value of n can vary greatly from one instance to another. This is due to factors such as the type of antenna used, the type of objects present, the layout of the environment, measurement system capabilities and/or uncertainties, and the height of the TX and RX, all of which could lead to different parameter values.

Secondly, in general, it can be seen that n increases as the directivity of the antenna decreases at either end or at both ends of the link. This is justified since the multipath effects become more prominent as the antenna directivity decreases. Similarly, n increases as the size of the room increases. Larger room size tends to lengthen delay spread. A general relationship can also be observed between n and $PL(d_0)$ in which higher n is accompanied by lower $PL(d_0)$. In addition, the small value of n in some LOS cases is compensated by a large $PL(d_0)$ value. The wide range of $PL(d_0)$ might be due to antenna misalignment or inaccurate antenna gain calibration.

Thirdly, [41] shows that antenna height has a strong influence on the value of n. As the antenna height increases, n decreases rather dramatically due to the establishment of a clear LOS path between the TX and RX.

Finally, [33] shows that under a similar measurement setup, the PL increases very rapidly in the case of cross-polar signal transmission as compared to the co-polar signal case. The PL of co-polar signal transmission can be well approximated by free space PL.

2.3.1.2 Shadowing

Shadowing signifies the average signal power received over a large area (a few tens of wavelengths) due to the dynamic evolution of propagation paths, whereby new paths arise and old paths disappear. Due to the variation in the surrounding environment, the received power will be different from the mean value for a given distance, which causes the PL variation about the mean value given in Equation (2.2) or (2.3).

Many measurement results reported in the 60 GHz range have shown that the shadowing fading is log-normally distributed [30, 42–44], that is, $X_\sigma[\text{dB}] = N(0, \sigma_S)$, where X_σ denotes a zero-mean Gaussian random variable measured in decibels with standard deviation σ_S. The value of σ_S is site-specific, as listed in Table 2.1 for different environments. The value of σ_S is higher at 60 GHz compared to lower frequencies since the transmission loss at 60 GHz is much higher than at lower frequencies.

The shadowing parameters derived here assume that the channel is static and there is no human movement. In the presence of human movement, measurement results show that the obstruction by humans can be significant and range from 18 to 36 dB [45]. Furthermore, the shadowing effect duration is relatively long, up to several hundred milliseconds, and increases with the number of persons in the environment [45].

2.3.2 Small-Scale Channel Characterization

Small-scale fading (i.e. fast fading) is caused by the multipath signals that arrive at the receiver with random phases that add constructively or destructively. It causes rapid changes in signal amplitude over a small distance (less than 10 wavelengths). Over this small local area, the small-scale fading is approximately superimposed on the constant large-scale fading. Before studying the various small-scale channel effects, we look at the generic channel model representation for the 60 GHz channel.

2.3.2.1 Generic Channel Model

Based on the clustering phenomenon observed in both the temporal and spatial domains from measurement data available in the literature such as [2, 29, 46, 47] and the references therein, a generic 60 GHz channel model that takes clustering into account is reasonable choice since it can always be reduced to the conventional single cluster channel model as observed in [35, 48, 49]. The proposed cluster model is based on the extension of the Saleh–Valenzuela (SV) model [50] to the angular domain by Spencer et al. [51]. The complex baseband directional CIR is given by

$$h(t, \phi, \theta) = \sum_{l=0}^{L} \sum_{k=0}^{K_l} \alpha_{k,l} \delta(t - T_l - \tau_{k,l}) \delta(\phi - \Omega_l - \omega_{k,l}) \delta(\theta - \Psi_l - \psi_{k,l}), \quad (2.4)$$

where $\delta(\cdot)$ is the Dirac delta function, L is the total number of clusters and K_l is the total number of rays in the lth cluster. The scalars $\alpha_{k,l}$, $\tau_{k,l}$, $\omega_{k,l}$ and $\psi_{k,l}$ denote the complex amplitude, time of arrival (ToA), angle of arrival (AoA) and angle of departure (AoD) of the kth ray of the lth cluster, respectively. Note that the AoA and AoD can consist either of azimuth domain, elevation domain, or both domains. Similarly, the scalars T_l, Ω_l and Ψ_l represent the mean ToA, mean AoA and mean AoD of the of the lth cluster. The key assumption used in deriving Equation (2.4) is that the spatial and temporal domains are independent and thus uncorrelated. However, measurement results in [52] have shown a correlation between these two domains, which can be modeled by two joint pdfs. It is also important to note that each multipath in Equation (2.4) will experience distortions due to the frequency dependence of the scatterers [52]. Both the frequency dependence and correlation of spatial and temporal domains are so far not taking into consideration in any 60 GHz channel model due to lack of measurement results. For very high-speed transmission, the channel described in Equation (2.4) can be assumed to remain approximately static over tens to thousands of symbols. This is also called block fading channel. This is mainly due to the low speed (up to 1.5 m/s) movement in indoor environment which yields a maximum Doppler frequency, f_D, of 300 Hz. Invoking the relationship between coherence time, T_{coh}, and f_D, $T_{coh} = 9/16\,\pi f_D$ [8], yields a coherence time of approximately 0.6 ms. The coherence time calculation for this model is based on the assumption that the spectral broadening is due to uniform multipath arrival over $[0, 2\pi]$, which is usually valid for cellular systems but might be doubtful at 60 GHz due to vastly different antenna and propagation characteristics in both systems. Measurement results reported in [13] show that the coherence time where the correlation starts to fall below 0.5 is no greater than 50 ms for all the measurements scenarios in which people walked along a clear LOS path at a speed of 1.7 m/s. Such a coherence time is significantly larger than the symbol time and corresponds to at least a few super-frames per beacon interval in IEEE 802.15.3 and IEEE 802.11.

2.3.2.2 Number of Clusters

The number of clusters is an important parameter for the channel models considered here. Different definitions of clustering appear in the literature. Clustering is defined in this book as a group of rays arriving at approximately the same time and angle. Various methods have been derived for cluster identification, ranging from simple visual inspection [50, 51] to advanced signal processing such as kernel density estimation [52]. Different results have been obtained to describe the numbers of clusters for wideband systems. In [27] it was reported that the number of clusters is Poisson distributed and can be fully characterized by a mean number of clusters, $\overline{L}$. On the other hand, the available literature on 60 GHz reports diverse findings for this parameter. Analysis of the measurement results in [35, 36, 48, 49] indicates that only a single cluster can be observed. Furthermore, analysis of IEEE 802.15.3c measurement data for various environments and scenarios shows that $\overline{L}$ does not follow a specific distribution. However, the observed mean number of clusters can be calculated by visual inspection. Values range typically from 3 to 14. These different findings could be due to several factors such as measurement bandwidth and richness of scatterers (i.e. number of objects) in the environment. In cases where the environment under consideration has more furniture, a larger number of clusters would be expected as a result of superstructure (such as walls, furniture, computers and doors) [53]. Higher bandwidth provides higher resolution that can resolve more multipath components, which can cluster more easily as they propagate in the channel.

2.3.2.3 Arrival Times

Under the assumption that the delay and angular domains can be modeled independently, the ToA of the generic channel model described in Equation (2.4) relies on two sets of parameters, namely, inter-cluster parameters, $\{T_l, L\}$, that characterize the cluster and intra-cluster parameters, $\{K_l, \alpha_{k,l}, t_{k,l}\}$, that characterize the multipath components.

The cluster arrival and ray arrival time distributions are described by two Poisson processes. According to this model, cluster inter-arrival times and ray intra-arrival times are given by two independent exponential pdfs. In particular, the cluster arrival time for each cluster is an exponentially distributed random variable conditioned on the cluster arrival time of the previous cluster, which can be expressed as

$$p(T_l|T_{l-1}) = \Lambda[-\Lambda \exp(T_l - T_{l-1})], \quad \text{for } l > 0, \tag{2.5}$$

where Λ is the cluster arrival rate. Similarly, the ray arrival time for each ray is an exponentially distributed random variable conditioned on the ray arrival time of the previous ray given by

$$p(\tau_{k,l}|\tau_{k-1,l}) = \lambda[-\lambda \exp(\tau_{k,l}|\tau_{k-1,l})], \quad \text{for } k > 0, \tag{2.6}$$

where λ is the ray arrival rate. In the classical SV model, T_0 and $\tau_{0,l}$ are assumed to be zero and all arrival times are relative with respect to the delay of the first path.

2.3.2.4 Arrival Angles

By invoking the assumption that the delay and angular domains, as well as AoA and AoD, are independent, the AoA and AoD of the generic channel model described in Equation (2.4) rely on two sets of parameters, namely, inter-cluster parameters $\{\Omega_l, \Psi_l, L\}$, that characterize the cluster and intra-cluster parameters $\{K_l, \alpha_{k,l}, \omega_{k,l}, \psi_{k,l}\}$, that characterize the multipath components.

Based on the 60 GHz measurement results reported in [54–56], the distribution of the cluster mean AoA, Ω_l, conditioned on the first cluster mean AoA, Ω_0, can be described by a uniform distribution over $[0, 2\pi]$, that is,

$$p(\Omega_l|\Omega_0) = \frac{1}{2\pi}, \quad \text{for } l > 0. \tag{2.7}$$

Note that the cluster AoA represents the mean of all AoAs within the cluster. On the other hand, the ray AoAs within each cluster can be modeled either by zero-mean Gaussian [56] or zero-mean Laplacian distributions [54, 55] as given respectively by

$$p(\omega_{k,l}) = \frac{1}{\sqrt{2\pi}\sigma_\phi} \exp\left(-\frac{\omega_{k,l}^2}{2\sigma_\phi^2}\right) \tag{2.8}$$

and

$$p(\omega_{k,l}) = \frac{1}{\sqrt{2}\sigma_\phi} \exp\left(-\left|\frac{\sqrt{2}\,\omega_{k,l}}{\sigma_\phi}\right|\right), \tag{2.9}$$

where σ_ϕ is the standard deviation. UWB measurements and spatial modeling reported in [57] also proposed that Laplacian distribution can be used to model the AoA information in a wideband channel. Similarly, the AoD can also be modeled using the appropriate pdf but unfortunately no such 60 GHz measurement and modeling are available to date to characterize the AoD.

2.3.2.5 Power Spectrum

The power delay angular spectrum (PDAS) of a directional channel is given by

$$P(t, \phi, \theta) = E\left[|h(t, \phi, \theta)|^2\right], \tag{2.10}$$

where $E[\cdot]$ and $|\cdot|$ denote the expectation and absolute value, respectively. In general, this can be simplified by assuming the delay, AoA and AoD are independent of each other, which leads to

$$P(t, \phi, \theta) = P(t)P(\phi)P(\theta), \tag{2.11}$$

where $P(t)$, $P(\phi)$ and $P(\theta)$ are the power delay spectrum (PDS), power AoA spectrum (PAAS), and power AoD spectrum (PADS), respectively. The PADS remains unavailable in the literature due the difficulty of obtaining it by measurement. Thus, the focus here is on the PDS and PAAS.

Power Delay Spectrum

The PDS of a channel, more commonly called the power delay profile (PDP), is the average power of the channel as a function of the excess delay with respect to the first arrival path. A number of important parameters can be derived by analyzing the PDP. In particular, the mean excess delay, root mean squared (rms) delay spread, timing jitter and standard deviation [29] can be obtained from the PDP of the channel. While these parameters are useful for specific system design, the most commonly used parameter is the rms delay spread, which is a second moment of the PDP that statistically measures the time dispersion of a channel. The rms delay spread, denoted τ_c, is inversely proportional to the coherence bandwidth of a channel, B_c, which determines whether a system is narrowband or wideband with respect to its channel. A narrowband system occurs when τ_c is less than the symbol period, T_s (or when B_c is larger than the signal bandwidth) of the system. This results in a flat fading channel. The reverse is true for a wideband system when $\tau_c > T_s$ (or when B_c is less than the signal bandwidth), which results in a frequency selective channel. A flat fading channel reduces the received signal-to-noise ratio due to deep fading, while a frequency selective channel causes inter-symbol interference that leads to irreducible bit error rate (BER) performance. Hence, τ_c determines the maximum transmission data rate in the channel without equalization and serves as a key design parameter for both single carrier (SC) and orthogonal frequency division multiplexing systems, as described in Chapter 5. A list of the rms delay values reported in the literature is given in Table 2.2.

The rms delay spread strongly depends on several factors. Firstly, the rms delay spread increases as the size of the room and the density of objects present in the room increase. This can be explained by the fact that for larger rooms with more objects, it takes longer for the multipath to reach the RX, prolonging the rms delay spread. Furthermore, the type of objects present in the environment can significantly affect the rms delay spread value. For instance, the increment in rms delay spread becomes more pronounced in the presence of more reflective objects such as metal than absorptive objects such as wood.

Secondly, the rms delay spread decreases as the directivity of the TX and/or RX antennas increases. Note, however, that such reduction in rms delay spread is only noticeable when the TX and RX antenna patterns are aligned to the most significant AoD and AoA, respectively. This can be explained by the spatial filtering introduced by the directive antennas which can "remove" undesired paths by focusing on the direction with the main path. However, as the misalignment increases, the rms delay spread will start to increase even if directive antennas are used at TX and RX. Ray tracing results in [68] show that in the case of perfect alignment, rms delay spread decreased by more than 50% from 27 ns (for omnidirectional antennas at both TX and RX) to 12.5 ns (for 60° HPBW antennas at both TX and RX). Conversely, in case of misalignment at TX with HPBW of 30° and omni RX, the rms delay

Table 2.2 The rms delay spread and the shape of the PDS of the measurement results reported in the literature

Ref	Environment	Scenario	τ_c (ns)	PDS shape	Comments
[58]	$(4.6 \times 6.0 \times 3.0)\,m^3$ room with plasterboard and concrete	LOS (empty)	2.9	NA	• TX 3 dB aperture around 70° in horizontal and vertical planes
		LOS (furnished)	2.7		• RX 3 dB aperture around 10°
		NLOS (simulation)	4.0		• Both at 1.5 m height
[59]	Common room with wooden table and chair $(56 \times 10\,m)$ – 3 sides with concrete wall and one side with glass	LOS	4.89	Smulders's PDP model	• TX/RX omnidirectional antenna (120°) • Both at 1.6 m
	Workshop with heavy machines	LOS	7.81		
[60]	Typical office $(8 \times 10)\,m$ with brick/stone and plasterboard. Partitions, desks and PCs in the room. All circular polarization	LOS (center of the room)	7.5–55.0	Saleh-Valenzuela shape	• TX omni HPBW 120° at 2.6 m, RX HPBW 120° at 1.3 m
			5.0–17.5		• TX omni HPBW 120° at 2.6 m, RX HPBW 15° at 1.3 m
		LOS (edge of the room)	9.0–35.0		• TX omni HPBW 120° at 2.6 m, RX HPBW 120° at 1.3 m
			5.0–22.5		• TX omni HPBW 120° at 2.6 m, RX HPBW 15° at 1.3 m
[42]	Corridor	LOS	14.7	N/A	• TX, RX omni-biconical at 1.8 m
	Canteen	LOS	13.5		
	Office	LOS	5.2		
	Corridor	NLOS	7.5		
	Office	NLOS	7.5		
	Parking	NLOS	26.5		
[61]	Reception room $(24.3 \times 11.2 \times 4.5)\,m^3$	LOS is based on measurement and NLOS is obtained by removing the LOS path.	45.0	Smulders's PDP	• TX, RX 9dBi biconical horn
	Computer Room $(9.9 \times 8.7 \times 3.1)\,m^3$		42.0		

Table 2.2 (*continued*)

Ref	Environment	Scenario	τ_c (ns)	PDS shape	Comments
	Lecture Room (12.9 × 8.9 × 4.0) m^3		18.0		
	Lab room (11.3 × 7.3 × 3.1) m^3		29.0		
	Amphitheather (30 × 21 × 6 m)		30.0		
	Hall (43 × 41 × 7) m^3		55.0		
	Vax Room (33.5 × 32.2 × 3.1) m^3		55.0		
	Corridor (44.7 × 2.4 × 3.1) m^3		70.0		
[62]	Empty conference room 90 m^2 area and 2.6 m height in a modern office building	LOS (VV)	11.0	NA	• TX horn 3 dB beamwidth 60° • RX lens-horn 3 dB beamwidth 4.6° both 1.46 m
		LOS (HH)	10.0		
		LOS (RR)	5.2		
[63]	Empty room (13.5 × 7.8 × 2.6) m^3 with plasterboard and concrete wall (measurement) TX omnidirectional with 2.36 m height	LOS	1.1	NA	• RX 1.5 m height with narrow (3 dB beamwidth 5°) – lens horn
		LOS	4.7		• RX 1.5 m height with medium (3 dB beamwidth 10°) gain pyramidal horn
		LOS	13.6		• RX 1.5 m height with broad (3 dB beamwidth 60°)- feed horn
		LOS	18.1		• RX 1.5 m height with omni (halfwave dipole)
[64]	Meeting room (5.0 × 7.0) m^2	LOS	0.66	Exponential decaying	• TX waveguide, RX waveguide
		NLOS	1.10		• TX waveguide, RX waveguide
	Computer lab (5.1 × 7.1) m^2	LOS	0.42		• TX patch, RX 4 patches, linear polarization
		LOS	0.77		• TX patch, RX 16 patches, linear polarization
		LOS	0.70		• TX 4 patches, RX 4 patches, linear polarization
		LOS	0.25		• TX 4 patches, RX 4 patches, circular polarization
		LOS	0.42		• TX 4 patches, RX 16 patches, linear polarization
		LOS	0.61		• TX 4 patches, RX 16 patches, circular polarization

(*continued overleaf*)

Table 2.2   (*continued*)

Ref	Environment	Scenario	τ_c (ns)	PDS shape	Comments
[65]	Room (furnished) (12.8 × 6.9 × 2.6) m³	LOS	9		• TX horn (10 dBi gain), RX omni
	Corridor (windows) (41 × 1.9 × 2.7) m³	LOS	31.6	NA	• TX, RX horn (10 dBi with 3 dB beamwidth of 69° and 55° in vertical and horizontal planes, respectively. Both at 1.7 m height
	Corridor (no windows) (41.0 × 1.6 × 2.7) m³	LOS	31.7		
[48]	Private home (wood and plasterboard construction)	Mixture of LOS and NLOS	3.0	Exponential decaying with peak/direct ray at zero delay.	• TX, RX vertically polarized biconical horn antennas with omni radiation
	Office (8–12 m²) including cubicles and conference rooms		6.5		• Three investigated different configurations: – TX at ceiling and RX on desk
	Laboratory (reflective metallic equipment and walls)		8.5		– TX on wall close to ceiling and RX on desk, and – Both TX and RX on desk
[36]	Corridor	LOS	15	Exponential decaying	• TX horn (22.7 dB) at 1.25 m, RX horn (5 dB) at 1.23 m
	Hall	LOS	16		• TX horn (5 dB) at 1.67 m, RX horn (5 dB) at 1.66 m
		NLOS	22		
[35]	Laboratory	LOS	7.3	Exponential decaying with peak/direct ray at zero delay	• TX omni at 1.4 m height, RX omni at 1.4 m height
			13.8		• TX omni at 1.9m height, RX omni at 1.4 m height
			20.8		• TX omni at 2.4m height, RX omni at 1.4 m height
		NLOS	12.9	Exponential decaying	• TX omni at 1.4 m height, RX omni at 1.4 m height
			14.8		• TX omni at 1.9m height, RX omni at 1.4 m height
			21.0		• TX omni at 2.4m height, RX omni at 1.4 m height
		LOS	14.6	Exponential decaying with peak/direct ray at zero delay.	• TX fan beam (16.5 dB with 70° HPBW) at 2.5 m height, RX omni at 1.4 m height
			1.2		• TX fan beam (16.5 dB with 70° HPBW) at 2.5 m height, RX fan beam (16.5 dB with 70° HPBW) at 1.4 m height

Table 2.2 (*continued*)

Ref	Environment	Scenario	τ_c (ns)	PDS shape	Comments
			1.1		• TX fan beam (16.5 dB with 70° HPBW) at 2.5 m height, RX pencil beam (24.4 dB with 8.3° HPBW) at 1.4 m height
[29]	Hallway in Durham Hall ($102.0 \times 2.1 \times 4.3$) m³	LOS	25.6	NA	• NA TX open-ended waveguide with 6.7 dB
	Hallway in Whittermore ($54.7 \times 2.9 \times 4.3$) m³	LOS	19.2		• HPBW 90° azimuth and 125° elevation
	Room in Durham ($6.7 \times 5.9 \times 4.3$) m³	LOS	4.9		• RX horn antenna with 29 dB HPBW are 7° in azimuth and 5.6° in elevation
	Room in Whittermore ($8.4 \times 7.0 \times 4.3$) m³	LOS	18.7		
	Hall way to Room ($11.7 \times 5.1 \times 4.3$) m³.	NLOS NLOS	5.8 18.3		
	Room to Room ($11.7 \times 5.1 \times 4.3$) m³ and ($5.1 \times 4.3 \times 4.2$) m³				
[49]	Corridor I ($1.30 \times 13.07 \times 2.57$) m³	LOS	10.5	Exponential decaying	• TX open-ended waveguide (OWG) antenna with 6 dBi gain
			6.2		• Lens antenna with effective 120° visible azimuth range
	Corridor II ($3.07 \times 12.25 \times 2.57$) m³	LOS	8.1		• TX OWG antenna with 6 dBi gain
			15.7		• Lens antenna with effective 120° visible azimuth range
	Corridor III ($5.50 \times 12.25 \times 2.57$) m³	LOS	12.3		• TX OWG antenna with 6 dBi gain
			13.4		• Lens antenna with effective 120° visible azimuth range
[66]	Empty Room (12.4 m $\times 8.1$ m)	LOS NLOS	6.58 16.14	NA	• TX biconical horn with 6 dBi gain, RX monopole with 4 dBi gain
[67]	Rectangular room (20 m²)	LOS	10		• TX horn with 20 dBi gain, RX horn with 20 dBi gain
	Corridor (3×30) m²		10		
	Lab (15×30) m²		30		

(*continued overleaf*)

Table 2.2   (*continued*)

Ref	Environment	Scenario	τ_c (ns)	PDS shape	Comments
[68]	Office $(20 \times 20)\,\mathrm{m}^2$	LOS	27.0	NA	• TX and RX half wave omni-dipole
			16.0		• TX directional (HPBW = 60°) and RX half wave omni-dipole
			14.0		• TX directional (HPBW = 30°) and RX half wave omni-dipole
			12.5		• TX directional (HPBW = 60°) and RX directional (HPBW = 60°)
			10.0		• TX directional (HPBW = 30°) and RX directional (HPBW = 30°)

spread for 60% of the locations increased marginally (from 9 ns to 10 ns) when the misalignment is 10°, and dramatically (from 9 ns to 21 ns) when the misalignment was above 30°. Measurement results reported in [44] also show a similar conclusion in terms of rms delay spread decreasing as the antenna directivity increases in the case of perfect alignment up to 35° of misalignment.

Power Angular Spectrum

The power angular spectrum of a channel is the average power of the channel as a function of angular information. A MIMO channel measurement is reported in [69], but no modeling work has been done as we go to press. However, several 60 GHz measurement results on single-input multiple-output (SIMO) configurations were reported in [29, 35, 36, 44, 46, 54, 55, 70] but only limited AoA modeling has been carried out to characterize the AoA information. The AoA information – such as mean AoA, AoA associated with maximum power and rms AoA spread – provides important design parameters for multiple antenna systems. For instance, the mean AoA and maximum power AoA provide a preferred beamformed direction. On the other hand, the rms AoA spread provides a statistical measure of the angular dispersion of a channel, which determines the amount of diversity and multiplexing gain that can be achieved from the channel. A list of the AoA measurements reported in the literature with the corresponding PAAS shape and AoA spread is given in Table 2.3.

Table 2.3 The rms AoA spread and the shape of the PAAS of the measurement results reported in the literature

Ref	Environment	Scenario	rms AoA spread (°)	PAS Shape	Comments
[49]	Corridor	LOS	14.5	Laplacian	• Average result over the measurement locations • TX open waveguide and lens antenna • RX 1 × 4 uniform rectangular array with patch antennas
[54]	Offlice	LOS	102	Laplacian	• TX antenna always fixed (horn antenna with HPBW 30°) • RX antenna rotated from 0° to 360° in 5° steps (30° beamwidth)
			66		• TX antenna always fixed (horn antenna with HPBW 30°) • RX antenna rotated from 0° to 360° in 5° steps (30° beamwidth)
[55]	Office	NLOS	9.6	Laplacian	• TX antenna always fixed (omni) • RX antenna rotated from 0° to 360° in 5° steps (15° beamwidth) • Antenna height 1.1 m • Antenna separation 10 m
[70]	Desktop	LOS	14.4	Gaussian	• TX omni, RX antenna rotated from 0° to 360° in 4° steps (<5° HPBW ~21 dBi) • Antenna separation 2–8 m
			34.6	Laplacian	• TX antenna always fixed (horn antenna with HPBW 30°) • Rx antenna rotated from 0° to 360° in 5° steps (30° beamwidth)
			38.1		• TX antenna always fixed (horn antenna with HPBW 30°) • RX antenna rotated from 0° to 360° in 5° steps (30° beamwidth)

It is important that we distinguish the ToA statistics and the PDS as well as AoA statistics and PAAS. For example, the ToA represents the frequency of paths that arrive within a certain time window (bin) without considering its power as long as the paths are within the dynamic range of the system. The PDS, on the other hand, represents the arrival of paths with a certain power. It is important to note that a larger number of paths arriving within the same time bin would not necessarily contribute to more received power as observed in the PDS, especially in the LOS scenario due to the presence of a dominant component.

2.3.2.6 Small-Scale Fading Statistics

Different from conventional narrowband systems, which consist of vector summation of many irresolvable paths as a result of limited capability of the system measurement bandwidth, the amplitude fading distributions are typically Rayleigh and Rician for NLOS and LOS scenarios, respectively. In indoor wideband systems, such as 60 GHz and UWB, small-scale amplitude fading can follow different distributions depending on type of environment, measurement bandwidth and scenario under consideration. For example, the Rayleigh distribution is used to model LOS for library environments using a 1 GHz bandwidth measurement system [46]. The Rayleigh distribution is also used to model LOS for office, home and library environments using a measurement system with 5 GHz bandwidth [48]. Measurement results with 100 MHz bandwidth in the laboratory [71] show that in the LOS case, the amplitude is Rician distributed for all channel taps, while in NLOS case only the first tap follows Rician distribution and the rest of the taps can be modeled by a Rayleigh distribution. Measurement in different corridors using 1 GHz bandwidth shows the amplitude is Rician with different K-factor values [49]. Elsewhere in the literature, the Weibull distribution is used to model LOS/NLOS for residential environments using a 7 GHz bandwidth [53], and the log-normal distribution is used to model LOS/NLOS for office environments using a 2 GHz bandwidth [71]. The Nakagami distribution is used to model office LOS using a 500 MHz bandwidth [72]. As the bandwidth of the measurement system increases sufficiently, more multipath components can be resolved and thus the effect of small-scale fading is expected to become less extreme.

2.3.3 Polarization

The use of polarization in wireless communications has recently gained a lot of attention, especially at lower frequencies due to limited mobile terminal size. Despite 60 GHz being less constrained in terms of antenna spacing, polarization does offer advantages in multipath environments. Measurement results in [62] show that the use of circular polarization can effectively reduce the rms delay spread by about 50% over vertical and horizontal polarization. This can be explained by the fact that circular polarization suppresses odd-order (1st, 3rd, ...) reflection. Furthermore, second-order reflection paths typically undergo higher attenuation at 60 GHz and become insignificant when they arrive at the RX. Thus, circular polarization can be a good candidate for LOS operation. It is shown in [73] that polarization can be exploited to increase the capacity in LOS for dual-polarized MIMO systems, while introducing negligible capacity degradation in NLOS.

The modeling of polarization in a MIMO setup has been studied in [74]. In particular, the spatial-polarized channel matrix for an $N_{TX} \times N_{RX}$ dual-polarized sub-array [74] is given by

$$\mathbf{H}_{sp} = \mathbf{H}_{spatial} \otimes \mathbf{X}_p, \qquad (2.12)$$

where $\mathbf{H}_{spatial}$ is an $N_{TX}/2 \times N_{RX}/2$ matrix that characterizes the spatial information of the sub-arrays and $\mathbf{X}_p$ is the 2×2 dual polarized matrix given by [74],

$$\text{vec}(\mathbf{X}_{VH \to VH}^H) = \begin{bmatrix} 1 & \sqrt{\mu\chi}\vartheta^* & \sqrt{\chi}\sigma^* & \sqrt{\mu}\delta_1^* \\ \sqrt{\mu\chi}\vartheta & \mu\chi & \sqrt{\mu}\chi\delta_2^* & \mu\sqrt{\chi}\sigma^* \\ \sqrt{\chi}\sigma & \sqrt{\mu}\chi\delta_2 & \chi & \sqrt{\mu\chi}\vartheta^* \\ \sqrt{\mu}\delta_1 & \mu\sqrt{\chi}\sigma & \sqrt{\mu\chi}\vartheta & \mu \end{bmatrix}^{1/2} \text{vec}(\mathbf{X}_w^H), \qquad (2.13)$$

where μ is the co-polar imbalance, χ is the polarization discrimination (XPD), δ_1 is the correlation between the VV and the HH components, δ_2 is the correlation between the VH and the HV components, σ is the receive correlation coefficient and ϑ is the transmit correlation coefficient. The matrix $\mathbf{X}_w$ is given by

$$\mathbf{X}_w = \begin{bmatrix} \exp(\phi_1) & \exp(\phi_3) \\ \exp(\phi_2) & \exp(\phi_4) \end{bmatrix}, \qquad (2.14)$$

where the φ_k ($k = 1, 2, 3$ and 4) are uniformly distributed on $[0, 2\pi]$. While these results can be extended to the 60 GHz context, no 60 GHz measurement results are available to verify the applicability of such model at 60 GHz. It has been shown that the 60 GHz polarization characteristics differ significantly compared to lower frequencies (e.g. XPD value [71]).

A slightly different modeling approach was taken in [75], whereby both the polarization characteristics of antennas and propagation are taken into consideration. Using the proposed modeling approach in [75], the CIR of Equation (2.4) is modified as

$$h(t, \phi, \theta) = \sum_{l=0}^{L} \sum_{k=0}^{K_l} \mathbf{E}\alpha_{k,l}\delta(t - T_l - \tau_{k,l})\delta(\phi - \Omega_l - \omega_{k,l})\delta(\theta - \Psi_l - \psi_{k,l}),$$

$$(2.15)$$

where $\mathbf{E}$ is the polarization matrix that defines the polarization characteristics under LOS and NLOS conditions. For LOS and NLOS scenarios $\mathbf{E}$ is given by the (scaled) identity matrix and reflection order polarization matrix, respectively. In NLOS scenarios, $\mathbf{E}$ for the first order is given by [75]

$$\mathbf{H}_{ref\,1}$$
$$= \mathbf{\Psi}_{RX}\mathbf{R}\mathbf{\Psi}_{TX}$$
$$= \underbrace{\begin{bmatrix} \cos(\psi_{rx}) & \sin(\psi_{rx}) \\ -\sin(\psi_{rx}) & \cos(\psi_{rx}) \end{bmatrix}}_{\substack{\text{recalculation} \\ \text{of polarization} \\ \text{vector from the} \\ \text{plane of incidence basis} \\ \text{to RX coordinates}}} \times \underbrace{\begin{bmatrix} R_{\perp}(\alpha_{inc}) & \xi_1 \\ \xi_2 & R_{\parallel}(\alpha_{inc}) \end{bmatrix}}_{\substack{\text{reflection} \\ \text{matrix} \\ \text{R}}} \times \underbrace{\begin{bmatrix} \cos(\psi_{tx}) & \sin(\psi_{tx}) \\ -\sin(\psi_{tx}) & \cos(\psi_{tx}) \end{bmatrix}}_{\substack{\text{recalculation} \\ \text{of TX polarization} \\ \text{vector to the plane} \\ \text{of incidence basis}}},$$

$$(2.16)$$

where scalars ψ_{rx} and ψ_{tx} are the incidence basis for the polarization vector to the RX and TX coordinates, respectively. Scalars ξ_1 and ξ_2 denote the cross polarization coupling coefficients. Scalars α_{inc}, $R_{\perp}(\alpha_{inc})$ and $R_{\parallel}(\alpha_{inc})$ represent the incident angle of the signal to the reflection surface, and reflection coefficients for the perpendicular and parallel components of the electric field as a function of incident angle, respectively. The second-order reflection is given by [75]

$$\mathbf{H}_{ref\,2} = \mathbf{\Psi}_{RX} \times \underbrace{\begin{bmatrix} R_{\perp}(\alpha_{2inc}) & \xi_1 \\ \xi_2 & R_{\parallel}(\alpha_{2\ inc}) \end{bmatrix}}_{\text{2nd reflection}} \times \begin{bmatrix} \cos(\psi_p) & \sin(\psi_p) \\ -\sin(\psi_p) & \cos(\psi_p) \end{bmatrix}$$

$$\times \underbrace{\begin{bmatrix} R_{\perp}(\alpha_{1inc}) & \xi_1 \\ \xi_2 & R_{\parallel}(\alpha_{1inc}) \end{bmatrix}}_{\text{1st reflection}} \times \mathbf{\Psi}_{TX}, \qquad (2.17)$$

where α_{1inc} and α_{2inc} are the incident angle for first and second reflections, respectively. The scalar ψ_p denotes the rotation angle between the first and second incident plane. The reflection polarization matrices, $\mathbf{R}$, are based on the fact that polarization changes with respect to reflection that depends on incident angle and roughness of the surface. In addition, the above model assumes that polarization characteristics differ between clusters but remain the same for rays within the same cluster. The modeling approach of [75] is complex and thus a lot of assumptions need to be made in order to simplify the model implementation, which is outside the scope of this chapter. Interested readers are referred to [75].

2.4   Industry Standard Channel Models

In this section, two industry standard channel models, IEEE 802.15.3c and IEEE 802.11ad, will be discussed since they are the key models used for evaluation of 60 GHz communications system.

2.4.1   IEEE 802.15.3c

IEEE 802.15.3c channel models are mainly derived based on wideband measurement results conducted in office, residential, library and desktop environments. For each environment, LOS and NLOS scenarios are defined. Some NLOS scenarios are generated from their LOS counterparts by appropriately removing the LOS component from the model [2].

2.4.1.1   Large-Scale Characterization

The IEEE 802.15.3c PL model adopted the conventional way to model the average PL without specifying a loss incurred by a specific object as given in Equation (2.3). The simple PL model is given by

$$\overline{PL}(d)[dB] = \underbrace{PL(d_0)[dB]}_{\substack{PL \text{ at reference} \\ \text{distance}}} + \underbrace{10n \log_{10}\left(\frac{d}{d_0}\right)}_{\substack{PL \text{ exponent at} \\ \text{relative distance } d}}, \quad \text{for } d \geq d_0. \qquad (2.18)$$

IEEE 802.15.3c defines PL models for residential and office in both LOS and NLOS scenarios. The parameters of these PL models are given in Table 2.4. Note that these parameters were derived from measurements using different values of transmitter antenna gain, G_{TX}, and receiver antenna gain, G_{RX}. In order to remove the effects of both antenna gains, one can compensate or adjust the proposed value of the parameter $PL(d_0)$ by a factor of $G_{TX} + G_{RX}$ as suggested in [76]. For example, the parameters of the PL model in residential and office environment summarized here were derived by eliminating the effects of both the TX and RX gains.

Under such approximation, assuming 0 dBi for the TX and RX antennas allows users to use their own antenna gain for link budget analysis. However, the proposed approximation becomes inaccurate when highly directive antennas are employed since only a limited number of multipaths could have reached the antenna, and thus the value of the parameters n and σ_S will be different [76]. The shadowing effect is log-normally distributed as described in Section 2.3.1.2.

Table 2.4 The PL exponent, n, and standard deviation for shadowing, σ_S, defined in IEEE 802.15.3c [2]. Reproduced by permission of © 2007 IEEE

Environment	Scenario	n	PL_0	σ_S	Comment	Ref.
Residential	LOS	1.53	75.1	1.5	TX 72° HPBW, RX 60° HPBW	[76]
Residential	NLOS	2.44	86.0	6.2	TX 72° HPBW, RX 60° HPBW	[76]
Office	LOS	1.16	84.6	5.4	TX omni, RX horn (30° HPBW)	[33]
Office	NLOS	3.74	56.1	8.6	TX omni, RX horn (30° HPBW)	[33]

2.4.1.2 Small-Scale Characterization

IEEE 802.15.3c adopted the concept of generic channel modeling as described in Equation (2.4). However, some measurement results show that when directive antennas are used in the measurement, especially in the LOS scenario, there appears a distinct and strong LOS path on top of the clustering phenomena. This LOS path can be included by adding a LOS component to (2.4) as given by

$$h(t, \phi, \theta) = \beta \delta(\tau, \phi, \theta)$$

$$+ \sum_{l=0}^{L} \sum_{k=0}^{K_l} \alpha_{k,l} \delta(t - T_l - \tau_{k,l}) \delta(\phi - \Omega_l - \omega_{k,l}) \delta(\theta - \Psi_l - \psi_{k,l}),$$

$$(2.19)$$

where the first term, $\beta \delta(\tau, \phi, \theta)$, accounts for the gain of the strict LOS component (i.e. the multipath gain of the first arrival path) which can be found deterministically using ray tracing or a simple geometry based method or statistically. The second term on the right-hand side of Equation (2.19) is exactly described as in the case of extended directional SV model.

For example, in a desktop LOS scenario, a two-path response was observed for the LOS component due to reflection off the table. In this case, $\beta \delta(\tau, \phi, \theta)$ can be modeled statistically as

$$\beta[dB] = 20 \log_{10} \left[\frac{\mu_d}{d} \left| \sqrt{G_{t1}G_{r1}} + \sqrt{G_{t2}G_{r2}} \Gamma_0 \exp \left[j \frac{4\pi h_1 h_2}{\lambda d} \right] \right| \right] - PL_d(\mu_d),$$

$$(2.20)$$

where

$$PL_d(\mu_d) = 20 \log_{10} \left(\frac{4\pi d_0}{\lambda} \right) + A_{NLOS} + 10 n_d \log_{10} \left(\frac{d}{d_0} \right),$$

$$(2.21)$$

where λ, A_{NLOS}, μ_d, Γ_0, h_1 and h_2 are the wavelength, attenuation value for NLOS environments, mean distance, reflection coefficient, and height of the TX and RX, respectively. $G_{t1}, G_{t2}, G_{r1}, G_{r2}$ are the gain of the TX antenna for path 1 and path 2, and gain of the RX antenna for path 1 and path 2, respectively. Equation (2.20) becomes deterministic for all the channels considered when Γ_0 is set to zero, and becomes stochastic when Γ_0 is non-zero for the LOS desktop scenario.

The mean number of clusters, $\overline{L}$, as observed in IEEE 802.15.3c channel model ranges from 3 to 14 depending on the environment and scenario (i.e. LOS or NLOS) under consideration. The number of clusters is obtained from uniform distribution using the value $\overline{L}$ for each channel realization.

Even in the presence of strong LOS, as when using a directive antenna as described in Equation (2.19), the directional SV model remains valid except that T_0 and $\tau_{0,l}$ are no longer zero since the reference zero point has been moved. With proper normalization with respect to strong LOS, the modified model behaves in a similar way to the conventional SV model.

The IEEE 802.15.3c channel model is a SIMO model which only characterizes the AoA. The 15.3c Task Group (TG3c) adopted a uniform distribution over $[0,2\pi]$ for the cluster mean AoA, Ω_l, conditioned on the first cluster mean AoA, Ω_0, as described in Equation (2.7). On the other hand, the ray AoAs within each cluster can be modeled either by zero-mean Gaussian or zero-mean Laplacian distributions given by Equations (2.8) and (2.9), respectively. Note that due to the presence of a strong LOS component as indicated in (2.19), the values of Ω_0 and $\omega_{0,l}$ are no longer zero since the reference zero point has been moved to the direction arrival of the LOS component. With proper normalization with respect to the strong LOS, the modified model behaves in a similar way to the directional SV model.

Despite various amplitude distributions reported in the literature, the analysis of the TG3c measurement results for different measurement system bandwidths and environments found that the both cluster and ray amplitudes can be modeled by log-normal distribution given by

$$p_l(r) = \frac{1}{\sqrt{2\pi}\sigma_r r} \exp\left(-\frac{(\ln r - \mu_r)^2}{2\sigma_r^2}\right), \tag{2.22}$$

where $\mu_r = E[\ln r]$ and σ_r^2 are the mean and variance of the Gaussian $\ln r$, respectively.

2.4.1.3   Channel Parameterization

The complete channel parameters for the IEEE 802.15.3c channel model are given in Tables 2.5–2.8 for residential, office, library and desktop environments, respectively.

Table 2.5 Parameters for LOS and NLOS residential environment denoted as CM1 and CM2, respectively. CM2 shall be derived from CM1 by removing the LOS path [2]. Reproduced by permission of © 2007 IEEE

Residential	LOS (CM1)					NLOS (CM2)
	TX 360°, RX 15°	TX 60°, RX 15°	TX 30°, RX 15°	TX 15°, RX 15°	TX 360°, RX 15°	
Λ [1/ns]	0.191	0.194	0.144	0.045	0.210	N/A
λ [1/ns]	1.22	0.90	1.17	0.93	0.77	N/A
Γ[ns]	4.46	8.98	21.50	12.60	4.19	N/A
γ [ns]	6.25	9.17	4.35	4.98	1.07	N/A
σ_c [dB]	6.28	6.63	3.71	7.34	1.54	N/A
σ_r [dB]	13.00	9.83	7.31	6.11	1.26	N/A
σ_ϕ [degree]	49.8	119.0	46.2	107.0	8.3	N/A
L	9	11	8	4	4	N/A
Δk [dB]	18.8	17.4	11.9	4.6	N/A	N/A
$\Omega(d)$ [dB] (derived at 3 m)	−88.7	−108.0	−111.0	−110.7	N/A	N/A
n_d (path loss exponent)	2	2	2	2	N/A	N/A
A_{NLOS}	0	0	0	0	N/A	N/A

Table 2.6 Parameters for LOS and NLOS residential environment denoted as CM3 and CM4, respectively [2]. Reproduced by permission of © 2007 IEEE

Office	LOS (CM3)		NLOS (CM4)		
	TX 30°, RX 30°	TX 60°, RX 60°	TX 360°, RX 15°	TX 30°, RX 15°	Omni TX, RX 15°
Λ [1/ns]	0.041	0.027	0.032	0.028	0.070
λ [1/ns]	0.971	0.293	3.450	0.760	1.880
Γ[ns]	49.8	38.8	109.2	134.0	19.4
γ [ns]	45.2	64.9	67.9	59.0	0.42
σ_c [dB]	6.60	8.04	3.24	4.37	1.82
σ_r [dB]	11.30	7.95	5.54	6.66	1.88
σ_ϕ [degree]	102.0	66.4	60.2	22.2	9.1
L	6	5	5	5	6
Δk [dB]	21.9	11.4	19.0	19.2	N/A
$\Omega(d)$ [dB]	−3.27d − 85.8	−0.303d − 90.3	−109.0	−107.2	N/A
n_d	2.00	2.00	3.35	3.35	N/A
A_{NLOS}	0	0	5.56 at 3 m	5.56 at 3 m	N/A

Table 2.7 Parameters for LOS library environment denoted as CM5. CM6 can be derived from CM5 by simply removing the LOS path and thus no parameters were provided [2]. Reproduced by permission of © 2007 IEEE

Library	LOS (CM5)	NLOS (CM6)
Λ [1/ns]	0.25	N/A
λ [1/ns]	4.0	N/A
Γ [ns]	12	N/A
γ [ns]	7.0	N/A
σ_c [dB]	5.0	N/A
σ_r [dB]	6.0	N/A
σ_ϕ [degree]	10.0	N/A
$\overline{L}$	9	N/A
K_{LOS} [dB]	8	N/A

2.4.2 IEEE 802.11ad

The IEEE 802.11ad channel model is currently under development. Two types of model are being developed under the 11ad Task Group (TGad) framework depending on the modeling approach used. The first uses ray tracing to derive statistical models [3]. While this uncommon modeling approach is used to reduce the measurement effort but its accuracy is debatable. The second type of model relies on channel measurement [77] to derive a statistical channel model. Hence, IEEE 802.11ad is a mixture of ray tracing and statistical modeling approaches. As we go to press, some models are still work in progress and we focus on the models for conference room and living room environments.

2.4.2.1 Large-Scale Characterization

The PL models in IEEE 802.11ad for conference rooms are derived based on averaging over large instantaneous PL values (channel realizations) generated from the statistical channel model at different TX–RX separation. Furthermore, the PL models are developed by using the basic directional antenna model with HPBW from $10°$ to $60°$ or a beamforming algorithm that adjusts the transmit and receive beams along the cluster with maximum power. The general representation of the PL model is given by

$$\overline{PL}(d)[dB] = A_c + 20\log_{10}(f) + 10n\log_{10}(d), \qquad (2.23)$$

where A_c is constant value. The first two terms of Equation (2.23) can be treated as $PL(d_0)$ as in Equation (2.3), though the two PL modeling approaches are different. A summary of values for A_c, n and $PL(d_0)$ at $f = 60$ GHz is given in Table 2.9 for different environments and scenarios obtained in IEEE 802.11ad based on analysis of ray tracing.

Table 2.8 Parameters for LOS desktop environment denoted as CM7. CM8 shall be derived from CM7 by simply removing the LOS path and thus no parameters were provided [2]. Reproduced by permission of © 2007 IEEE

Desktop	LOS (CM7)		LOS (CM7)	NLOS
	TX 30°, RX 30°	TX 60°, RX 60°	Omni TX, RX 21 dBi	(CM8)
Λ [1/ns]	0.037	0.047	1.720	N/A
λ [1/ns]	0.641	0.373	3.140	N/A
Γ[ns]	21.10	22.30	4.01	N/A
γ [ns]	8.85	17.2	0.58	N/A
σ_c [dB]	3.01	7.27	2.70	N/A
σ_r [dB]	7.69	4.42	1.90	N/A
σ_ϕ [degree]	34.6	38.1	14.0	N/A
$\overline{L}$	3	3	14	N/A
Δk [dB]	11.0	17.2	N/A	N/A
$\Omega(d)$ [dB]	$4.44d - 105.4$	$3.46d - 98.4$	N/A	N/A
h_1	Uniform dist. Range: 0–0.3	Uniform dist. Range: 0–0.3	N/A	N/A
h_2	Uniform dist. Range: 0–0.3	Uniform dist. Range: 0–0.3	N/A	N/A
D	Uniform dist. Range: $d \pm 0.3$	Uniform dist. Range: $d \pm 0.3$	N/A	N/A
G_{T1}	GSS*	GSS	N/A	N/A
G_{R1}	GSS	GSS	N/A	N/A
G_{T2}	GSS	GSS	N/A	N/A
G_{R2}	GSS*	GSS	N/A	N/A
n_d	2	2	N/A	N/A
A_{NLOS}	0	0	N/A	N/A

*GSS is a Gaussian antenna model with sidelobe level discussed in [78].

2.4.2.2 Small-Scale Characterization

The IEEE 802.11ad model adopts the generic CIR introduced in IEEE 802.15.3c. However, TGad uses a simpler approach to model the LOS component described in Equation (2.19), whereby a Friis model is used to derive the amplitude of the LOS path. The LOS path is a cluster with a single ray path that can be computed as

$$\beta_{\text{LOS}}[dB] = G_t + G_r + 20 \log_{10}\left(\frac{\lambda}{4\pi d}\right). \tag{2.24}$$

Table 2.9 The PL exponent, n, and standard deviation for shadowing, σ_S, defined in IEEE 802.11ad

Environment	Scenario	A_c	n	$PL(d_0)$	σ_s	Ref.
Conference room STA-STA	LOS	32.5	2.0	68.0	0	[47]
Conference room STA-STA	NLOS	51.5	0.6	87.0	3.3	[47]
Conference room STA-AP	LOS	32.5	2.0	68.0	0	[47]
Conference room STA-AP	NLOS	45.5	1.4	81.1	3.0	[47]
Living room	LOS	32.5	2.0	68.0	0	[47]
Living Room	NLOS	44.7	1.5	80.2	3.4	[47]

On the other hand, the gain of a cluster, β_i, is given by

$$\beta_i[dB] = 20\log_{10}\left(\frac{g_i\lambda}{4\pi(d+R)}\right), \tag{2.25}$$

where g_i is a reflection loss, d is distance between TX and RX (along LOS path), R is a total distance along the cluster path decreased by d which is a product of ToA of the LOS and the speed of light. Note that g_i is modeled by a log-normal distribution with mean value $-10\,$dB and rms value $4\,$dB in the case of first-order reflection, and with mean value $-16\,$dB and rms value $5\,$dB in the case of second-order reflection.

Furthermore, the model also accounts for cluster blockage due to objects present in the conference room such as humans and laptops. For simplicity, each type of cluster is modeled using cluster blockage probability. The probability of blockage in the STA-STA scenario is given in Table 2.10.

The IEEE 802.11ad model assumes several types of clusters, with a fixed number of clusters for each cluster type. Four types of cluster can be identified, originating from the first-order reflection from walls, first-order reflection from ceiling, second-order reflection from walls and second-order reflection from ceiling, respectively. Depending on the scenario and environment, some clusters might be present or absent from the model. The numbers of clusters for LOS and NLOS in the conference room and living room scenarios are shown in Table 2.11. The results obtained in IEEE 802.11ad model are derived from ray tracing simulation by placing the AP and STA in predefined locations [3]. For each channel realization, the fixed number of clusters is used to generate the directional CIR.

The IEEE 802.11ad model obtains the empirical distribution of the ToA of the different cluster types through ray tracing simulation. In the conference room environment with the STA-STA scenario, the ToAs of the cluster arising due to the first- and second-order reflection from wall and ceiling, respectively, are modeled by four inter-cluster ToA pdfs shown in Table 2.12.

Table 2.10 Probabilities of clusters blockage for STA-STA and STA-AP scenarios [3]. Reproduced by permission of © 2010 IEEE

Scenario	Cluster type	Probability of cluster blockage
Conference room STA-STA	LOS	0 or 1 (set as a model parameter)
	First-order reflections from walls	0.24
	First-order reflections from ceiling	0
	Second-order reflections from wall and ceiling	0.037
	Second-order reflections from walls	$p = 0.175$ (binomial distribution parameter)
Conference room STA-AP	LOS	0 or 1 (set as a model parameter)
	First-order reflections from walls	0.126
	Second-order reflections from walls	$p = 0.07$ (binomial distribution parameter)

Table 2.11 Number of clusters for each cluster groups in the STA-STA and STA-AP scenarios for conference room and living room environments [3]. Reproduced by permission of © 2010 IEEE

Environment	Type of clusters	# of clusters for STA-STA scenario	# of clusters for STA-AP scenario
Conference room	LOS path	1	1
	1st-order reflections from walls	4	4
	2nd-order reflections from two walls	8	8
	1st-order reflection from ceiling	1	-
	2nd-order reflections from walls and ceiling	4	-
Living room	LOS path	1	NA
	1st-order reflections from walls	3	NA
	2nd-order reflections from two walls	5	NA
	1st-order reflection from ceiling and floor	2	NA
	2nd-order reflections from the walls and floor	2	NA
	2nd-order reflections from the ceiling, walls and floor	6	NA

Table 2.12 Inter-cluster ToA pdf for first- and second-order reflection in conference room and living room environment

Environment and scenario	Description	Inter-cluster ToA pdf
Conference room STA-STA	1st-order reflection from wall	$W_t(t) = \begin{cases} 0, & t < 4 \\ 0.0577 \cdot t - 0.2307, & 4 \leq t < 7 \\ -0.0307 \cdot t + 0.3882, & 7 \leq t < 11 \\ -0.0042 \cdot t + 0.0958, & 11 \leq t < 23 \\ 0, & t \geq 23 \end{cases}$
	2nd-order reflection from walls	$W_t(t) = \begin{cases} 0, & t < 10 \\ 0.08, & 10 \leq t < 20 \\ 0.02, & 20 \leq t < 30 \\ 0, & t \geq 30 \end{cases}$
	1st-order reflection from ceiling	$W_t(t) = \begin{cases} 0, & t < 7 \\ 0.0677 \cdot t - 0.4741, & 7 \leq t < 11 \\ -0.0797 \cdot t + 1.1473, & 11 \leq t < 14.4 \\ 0, & t \geq 14.4 \end{cases}$
	2nd-order reflection from walls and ceiling	$W_t(t) = \begin{cases} 0, & t < 10.5 \\ 0.0551 \cdot t - 0.5790, & 10.5 \leq t < 14 \\ -0.0358 \cdot t + 0.6935, & 14 \leq t < 18 \\ -0.0071 \cdot t + 0.1786, & 18 \leq t < 25 \\ 0, & t \geq 25 \end{cases}$
Conference room STA-AP	1st-order reflection from wall	$W_t(t) = \begin{cases} 0, & t < 1 \\ 0.0977 \cdot t - 0.0983, & 1 \leq t < 3 \\ -0.0133 \cdot t + 0.2370, & 3 \leq t < 5 \\ -0.0760 \cdot t + 0.5507, & 5 \leq t < 7 \\ 0.0003 \cdot t + 0.0162, & 7 \leq t < 18 \\ -0.0110 \cdot t + 0.2237, & 18 \leq t < 20 \\ 0, & t \geq 20 \end{cases}$
	2nd-order reflection from wall	$W_t(t) = \begin{cases} 0, & t < 4 \\ 0.0535 \cdot t - 0.2170, & 4 \leq t < 6 \\ -0.0527 \cdot t + 0.4247, & 6 \leq t < 8 \\ 0.0194 \cdot t - 0.1520, & 8 \leq t < 13 \\ -0.0225 \cdot t + 0.3921, & 13 \leq t < 16 \\ 0.0003 \cdot t + 0.0271, & 16 \leq t < 20 \\ -0.0113 \cdot t + 0.2600, & 20 \leq t < 23 \\ 0, & 23 \leq t < 25 \\ 0.0200 \cdot t - 0.5005, & 25 \leq t < 28 \\ -0.0300 \cdot t + 0.9009, & 28 \leq t < 30 \\ 0, & t \geq 30 \end{cases}$

(continued overleaf)

Table 2.12   (*continued*)

Environment and scenario	Description	Inter-cluster ToA pdf
Living room STA-STA	1st-order reflection from walls	$W_t(t) = \begin{cases} 2/89 & t < 9 \\ 4/89 & 9 \le t < 23 \\ 1/89 & 23 \le t < 38 \\ 0 & t \ge 38 \end{cases}$
	2nd-order reflection from walls	$W_t(t) = \begin{cases} 0 & t < 9 \\ 1/114 & 9 \le t < 23 \\ 4/114 & 23 \le t < 48 \\ 0 & t \ge 48 \end{cases}$
	1st-order reflection from ceiling and floor	$W_t(t) = \begin{cases} 0 & t < 1 \\ 0.1667 \cdot t - 0.1667 & 1 \le t < 3 \\ -0.0833 \cdot t + 0.5833 & 3 \le t < 7 \\ 0 & t \ge 7 \end{cases}$
	2nd-order reflection from ceiling and floor	$W_t(t) = \begin{cases} 0 & t < 7 \\ 1/9 & 7 \le t < 16 \\ 0 & t \ge 16 \end{cases}$
	2nd-order reflection from wall, ceiling and floor	$W_t(t) = \begin{cases} 0 & t < 2 \\ 2/89 & 2 \le t < 10 \\ 4/89 & 10 \le t < 25 \\ 1/89 & 25 \le t < 38 \\ 0 & t \ge 38 \end{cases}$

On the other hand, for the intra-cluster case, three types of rays are specified, namely, a central ray, pre-cursor rays and post-cursor rays. A central ray, $\alpha_{i,0}$, is a single ray with fixed amplitude, while pre-cursor rays, and post-cursor rays are a group of rays with total number of rays given by N_{pre} and N_{post}, respectively. Both pre-cursor and post-cursor ray types are modeled using the Poisson process given by Equation (2.6) with arrival rate, λ_{pre} and λ_{post}, respectively. Furthermore, the average amplitudes at delay τ of pre-cursor rays, $A_{pre}(\tau)$, and post-cursor rays, $A_{post}(\tau)$, are exponentially decaying at rate γ_{pre} and γ_{post}, respectively.

Table 2.13 Intra-cluster parameters for conference room and living room environment [3]. Reproduced by permission of © 2010 IEEE

Parameter	Conference room	Living room
Pre-cursor rays K-factor, K_{pre}	10 dB	11.5 dB
Pre-cursor rays power decay time, γ_{pre}	3.7 ns	1.25 ns
Pre-cursor arrival rate, λ_{pre}	0.37 ns^{-1}	0.28 ns^{-1}
Pre-cursor rays amplitude distribution	Rayleigh	Rayleigh
Number of pre-cursor rays, N_{post}	6	6
Post-cursor rays K-factor, K_b	14.2 dB	10.9 dB
Post-cursor rays power decay time, γ_b	4.5 ns	8.7 ns
Post-cursor arrival rate, λ_b	0.31 ns^{-1}	1.0 ns^{-1}
Post-cursor rays amplitude distribution	Rayleigh	Rayleigh
Number of post-cursor rays, N_b	8	8

A_{pre} and A_{post} are related to $\alpha_{i,0}$ by

$$K_{pre} = 20 \log 10 \left| \frac{\alpha_{i,0}}{A_{pre}(\tau=0)} \right|, \tag{2.26}$$

$$K_{post} = 20 \log 10 \left| \frac{\alpha_{i,0}}{A_{post}(\tau=0)} \right|, \tag{2.27}$$

where K_{pre} and K_{post} are the K-factor values for the pre-cursor ray and post-cursor ray cases, respectively. Table 2.13 shows the values for the related intra-cluster parameters in conference room and living room environments.

Unlike the IEEE 802.15.3c model, the IEEE 802.11ad model is a MIMO model. In the IEEE 802.11ad model, four types of angular information are characterized corresponding to the first-order reflection from the wall, first-order reflection from the ceiling, second-order reflection from the wall and second-order reflection from the ceiling. The generation of the cluster AoA and AoD for the given type of angular information follows specific properties as described in [3]. The AoA, AoD and elevation-angle of arrival (EoA) distributions for STA-STA and STA-AP scenario are given in Table 2.14

In the IEEE 802.11ad model, the pre-cursor and post-curser rays have uniformly distributed phases over $[0, 2\pi]$ and Rayleigh distributed amplitudes with mean values as described in [3]. The amplitudes of the first pre-cursor and post-curser ray are relative to the central ray by a factor given by 5 dB and 10 dB, respectively.

Table 2.14 Inter-cluster AoA and AoD pdfs for first- and second-order reflection in conference room and living room environments

Environment/ scenario	Description	AoA or EoA pdf
Conference room STA-STA	1st-order reflection from wall	$\begin{cases} \varphi_B \geq \left(\dfrac{140}{90}\right) \cdot \varphi_A + 62° \\ \varphi_B \leq \left(\dfrac{140}{90}\right) \cdot \varphi_A + 82° \\ \varphi_B \leq 180° \\ \varphi_A \geq 0° \end{cases}$, $\begin{cases} \varphi_B \leq \left(\dfrac{140}{90}\right) \cdot \varphi_A - 62° \\ \varphi_B \geq \left(\dfrac{140}{90}\right) \cdot \varphi_A - 82° \\ \varphi_B \geq -180° \\ \varphi_A \leq 0° \end{cases}$. where φ_A and φ_B are the pair of azimuth angles for cluster A and B, respectively
	2nd-order reflection from wall	$\begin{cases} \varphi_{tx1} = 180° \cdot u_1 \\ \varphi_{rx1} = \varphi_{tx1} \end{cases}$, $\begin{cases} \varphi_{tx2} = 180° \cdot u_2 \\ \varphi_{rx2} = \varphi_{tx2} \end{cases}$, $\begin{cases} \varphi_{tx3} = 180° \cdot (u_3 - 1) \\ \varphi_{rx3} = \varphi_{tx3} \end{cases}$ $\begin{cases} \varphi_{tx4} = 180° \cdot (u_4 - 1) \\ \varphi_{rx4} = \varphi_{tx4} \end{cases}$, $\begin{cases} \varphi_{tx5} = 180° \cdot (u_5 - 1) \\ \varphi_{rx5} = \varphi_{tx5} + 180° \end{cases}$, $\begin{cases} \varphi_{tx6} = 180° \cdot (u_6 - 1) \\ \varphi_{rx6} = \varphi_{tx6} + 180° \end{cases}$, $\begin{cases} \varphi_{tx7} = 180° \cdot u_7 \\ \varphi_{rx7} = \varphi_{tx7} - 180° \end{cases}$, $\begin{cases} \varphi_{tx8} = 180° \cdot u_8 \\ \varphi_{rx8} = \varphi_{tx8} - 180° \end{cases}$, where $\varphi_{tx1}, \varphi_{rx1}, \dots, \varphi_{tx8}, \varphi_{rx8}$ are the TX and RX azimuth angle pairs, and $u_1, \dots, u_8$ are uniformly distributed on $[0,1]$
	1st-order reflection from ceiling	$W_\theta(\theta) = \begin{cases} 0 & , \quad \theta < 56.6° \\ 0.0023 \cdot \theta - 0.1302 & , \quad 56.6° \leq \theta < 83° \\ -0.0087 \cdot \theta + 0.7810 & , \quad 83° \leq \theta < 90° \\ 0 & , \quad \theta \geq 90° \end{cases}$ where θ is the elevation angle
	2nd-order reflection from wall and ceiling	$W_\theta(\theta) = \begin{cases} 0 & , \quad \theta < 30° \\ 0.0028 \cdot \theta - 0.0833 & , \quad 30° \leq \theta < 52° \\ -0.0056 \cdot \theta + 0.3500 & , \quad 52° \leq \theta < 63° \\ 0 & , \quad \theta \geq 63° \end{cases}$

Table 2.14 (*continued*)

Environment/ scenario	Description	AoA or EoA pdf
Conference room STA-AP	1st-order reflection from wall	$\begin{cases} \varphi_B \geq \left(\dfrac{135}{80}\right) \cdot \varphi_A + 45° \\ \varphi_B \leq \left(\dfrac{135}{80}\right) \cdot \varphi_A + 80°, \\ \varphi_B \leq 180° \\ \varphi_A \geq 0° \end{cases}$ $\begin{cases} \varphi_B \leq \left(\dfrac{135}{80}\right) \cdot \varphi_A - 45° \\ \varphi_B \geq \left(\dfrac{135}{80}\right) \cdot \varphi_A - 80° \\ \varphi_B \geq -180° \\ \varphi_A \leq 0° \end{cases}$
	2nd-order reflection from wall	Same as the pdf defined in STA-STA 2nd-order reflection from wall
	LOS path for elevation angle	$W_\theta(\theta) = \begin{cases} 0, & \theta < -76° \\ 1/45, & -76° \leq \theta < -31° \\ 0, & \theta \geq -31° \end{cases}$
	1st-order reflection	$W_\theta(\theta) = \begin{cases} 0, & \theta < -52° \\ 1/132, & -52° \leq \theta < -37° \\ 1/22, & -37° \leq \theta < -22° \\ 1/44, & -22° \leq \theta < -13° \\ 0, & \theta \geq -13° \end{cases}$
	2nd-order reflection	$W_\theta(\theta) = \begin{cases} 0, & \theta < -34° \\ 3/154, & \theta 34° \leq \theta < -20 \\ 6/77, & -20° \leq \theta < -12° \\ 2/77, & -12° \leq \theta < -8° \\ 0, & \theta \geq -8° \end{cases}$
Living room STA-STA	1st-order reflection from walls	$\begin{cases} \varphi_B \geq \left(\dfrac{82}{62}\right) \cdot \varphi_A + \left(\dfrac{2423}{31}\right)° \\ \varphi_B \leq \left(\dfrac{82}{37}\right) \cdot \varphi_A + \left(\dfrac{2396}{37}\right)°, \\ \varphi_B \leq 180° \end{cases}$ $\begin{cases} \varphi_B \leq \left(\dfrac{82}{62}\right) \cdot \varphi_A - \left(\dfrac{2423}{31}\right)° \\ \varphi_B \geq \left(\dfrac{82}{37}\right) \cdot \varphi_A - \left(\dfrac{2396}{37}\right)°, \\ \varphi_B \geq -180° \end{cases}$

(*continued overleaf*)

Table 2.14   (*continued*)

Environment/ scenario	Description	AoA or EoA pdf

$$\begin{cases} \varphi_B \geq \left(\dfrac{82}{52}\right) \cdot \varphi_A + 60° \\ \varphi_B \leq \left(\dfrac{82}{52}\right) \cdot \varphi_A + 75° \\ \varphi_B \leq 150° \\ \varphi_A \geq 0° \end{cases}, \quad \begin{cases} \varphi_B \leq \left(\dfrac{82}{52}\right) \cdot \varphi_A - 60° \\ \varphi_B \geq \left(\dfrac{82}{52}\right) \cdot \varphi_A - 75° \\ \varphi_B \geq -150° \\ \varphi_A \leq 0° \end{cases}$$

2nd-order reflection from walls

$$W_\varphi(\varphi) = \begin{cases} 0 & \varphi < -180° \\ 2/423 & -180° \leq \varphi < -129° \\ 0 & -129° \leq \varphi < 73° \\ 3/423 & -129° \leq \varphi < 73° \\ 0 & \varphi > 180° \end{cases} \quad \text{for walls 1–4,}$$

$$W_\varphi(\varphi) = \begin{cases} 0 & \varphi < -180° \\ 3/423 & -180° \leq \varphi < -73° \\ 0 & -73° \leq \varphi < 129° \\ 2/423 & 129° \leq \varphi \leq 180° \\ 0 & \varphi > 180° \end{cases} \quad \text{for walls 1, 2,}$$

$$W_\varphi(\varphi) = \begin{cases} 0 & \varphi < 18° \\ 1/130 & 18° \leq \varphi < 148° \\ 0 & \varphi \geq 148° \end{cases} \quad \text{for walls 4, 2,}$$

$$W_\varphi(\varphi) = \begin{cases} 0 & \varphi < -148° \\ 1/130 & -148° \leq \varphi < -18° \\ 0 & \varphi \geq -18° \end{cases} \quad \text{for walls 2, 4,}$$

$$W_\varphi(\varphi) = \begin{cases} 0 & \varphi < -54° \\ 1/108 & -54° \leq \varphi < 54° \\ 0 & \varphi \geq 54° \end{cases} \quad \text{for walls 3, 1}$$

1st-order reflection from ceiling and floor

$$W_\theta(\theta) = \begin{cases} 0 & \theta < 25° \\ 0.01320 \cdot \theta - 0.33000 & 25° \leq \theta < 29° \\ -0.00155 \cdot \theta + 0.09761 & 29° \leq \theta < 62.5° \\ 0 & \theta \geq 62.5° \end{cases}$$

2nd-order reflection from ceiling and floor

$$W_\theta(\theta) = \begin{cases} 0 & \theta < 25° \\ 0.01320 \cdot \theta - 0.33000 & 25° \leq \theta < 29° \\ -0.00155 \cdot \theta + 0.09761 & 29° \leq \theta < 62.5° \\ 0 & \theta \geq 62.5° \end{cases}$$

Table 2.14 (*continued*)

Environment/ scenario	Description	AoA or EoA pdf

2nd-order reflection from walls and ceiling

$$W_\theta(\theta) = \begin{cases} 0 & \theta < 12° \\ 1/8 & 12° \leq \theta < 20° \\ 0 & \theta \geq 20° \end{cases} \quad \text{for wall 1 to ceiling}$$

$$W_\theta(\theta) = \begin{cases} 0 & \theta < 13° \\ 1/11 & 13° \leq \theta < 24° \\ 0 & \theta < 24° \end{cases} \quad \text{for ceiling to wall 2/4}$$

$$W_\theta(\theta) = \begin{cases} 0 & \theta < 18° \\ 1/21 & 18° \leq \theta < 39° \\ 0 & \theta \geq 39° \end{cases} \quad \text{for wall 2/4 to ceiling}$$

2.5 Summary

In this chapter, we have presented a review of 60 GHz channel modeling, in which the small-scale and large-scale channel characterizations are thoroughly discussed. In particular, the PL and shadowing effects impose huge losses to the communication link. For example, at distance of 10 m, a total of 110.4 dB loss is incurred by the path loss in NLOS in the residential environment in IEEE 802.15.3c channel model (see Table 2.4). These losses need to be compensated in order to deliver the promise of multi-gigabit wireless transmission. By using beamforming technology, one can easily provide steerable 15 dB gain over the space of interest at one end of the link. Beamforming is very critical to 60 GHz communications (see Chapter 5). Despite the availability of the diverse channel models in the literature, a true 60 GHz MIMO channel model based on measurement results is still lacking. More such measurements and modeling work should be conducted to further characterize the MIMO propagation mechanisms in the 60 GHz band. Furthermore, previous results have demonstrated the applicability of circularly polarized signals for 60 GHz transmission; however, a generic channel model for different polarization signals is still unavailable and certainly deserves more in-depth investigation.

References

[1] Erceg, V. (2004) TGn Channel Models. IEEE 802.11-03-940-04, May.
[2] Yong, S.K (2007) TG3c Channel Modeling Sub-committee Final Report IEEE 802.15-07-0584-01-003c, March.
[3] Maltsev, A. (2009) Channel models for 60 GHz WLAN systems. IEEE 802.11-09-0334-03ad, July.

[4] 3rd Generation Partnership Project, Technical Specification Group Radio Access Network (2007) Spatial channel model for Multiple Input Multiple Output (MIMO) simulations (Release 7.0). 3GPP TR 25.996 V7.0.0, June.

[5] Steinbauer, M., Molisch, A.F and Bonek, E. (2001) The double-directional radio channel. *IEEE Communications Magazine*, **43**(4).

[6] Fleury, B.H and Leuthold, P.E (1996) Radiowave propagation in mobile communications: an overview of European research. *IEEE Communications Magazine*, **34**(2), 70–81.

[7] Saunders, S.R (1999) *Antennas and Propagation for Wireless Communication Systems*. Chichester: John Wiley & Sons, Ltd..

[8] Rappaport, T.S (2002) *Wireless Communications: Principles and Practice*, 2nd edn. Upper Saddle River, NJ: Prentice Hall PTR.

[9] Vaughan, R. and Bach Andersen, J. (2003) *Channels, Propagation and Antennas for Mobile Communications*. London: IEE Press.

[10] Janaswamy, R. (2006) An indoor path loss based on 60 GHz transport theory. *IEEE Antennas and Wireless Propagation Letters*, **5**(1), 58–60.

[11] Smulders, P. and Correia, L. (1997) Characterisation of propagation in 60 GHz radio channels. *Electronic and Communication Engineering Journal*, **9**(2), 73–80.

[12] Hansen, J. (2002) A novel stochastic millimeter-wave indoor radio channel. *IEEE Journal on Selected Areas in Communications*, **20**(6), 1240–1246.

[13] Moraitis, N. and Constantinou, P. (2002) Indoor channel modeling at 60 GHz for wireless LAN applications. *Proceedings of the IEEE International Symposium on Personal, Indoor and Mobile Radio Communication*, vol. 3, pp. 1203–1207, September.

[14] Lostalen, Y., Corre, Y., Louët, Y., Helloco, Y.L, Collonge, S. and Zein, G.E (2002) Comparison of measurements and simulations in indoor environments for wireless local area networks. *Procddings of the IEEE Vehicular Technology Conf*erence, vol. 1, pp. 389–393, May.

[15] Rappaport, T.S and Sandhu, S. (1994) Radio-wave propagation for emerging wireless personal-communication systems. *IEEE Antennas and Propagation Magazine*, **36**(5), 14–24.

[16] Peter, M., Keusgen, W. and Felbecker, R. (2007) Measurement and ray-tracing simulation of the 60 GHz indoor broadband channel: model accuracy and parameterization. *2nd European Conference on Antennas and Propagataion*, pp. 1–8, November.

[17] Khafaji, A., Saadane, R. El Abbadi, J. and Belkasmi, M. (2008) Ray tracing technique based 60 GHz band propagation modelling and influence of people shadowing. *International Journal of Electrical, Computer, and Systems Engineering*, pp. 102–108.

[18] Jacob, M., Kumer, T. and Chambelin, P. (2009) Deterministic channel modeling for 60 GHz WLAN. IEEE 802.11-09-302-00-ad, July.

[19] Lim, C.-P., Lee, M. Burkholder, R.J, Volakis, J.L Marhefka, R.J (2009) 60 GHz indoor propagation studies for wireless communications based on a ray-tracing method. *EURASIP Journal on Wireless Communications and Networking*, article ID 73928, January.

[20] Yong, S.K, Chong, C.C and Lee, S.S (2005) General guidelines for measurement techniques and procedures. IEEE 802.15-05-0357-00-003c, July.

[21] Liberti, J.C and Rappaport, T.S (1996) A geometrically based model for line-of-sight multipath radio channels. *Proceedings of the IEEE Vehicular Technology Conference*, pp. 844–848, April.

[22] Petrus, P., Reed, J.H and Rappaport, T.S (2002) Geometrical-based statistical macrocell channel model for mobile environments. *IEEE Transactions on Communications*, **50**(3), 495–502.

[23] Ertel, R.B and Reed, J.H (1999) Angle and time of arrival statistics for circular and elliptical scattering models. *IEEE Journal on Selected Areas in Communications*, **17**(11), 1829–1840.

[24] Svantesson, T. (2002) A double-bounce channel model for multi-polarized MIMO systems. *Proceedings of the IEEE Vehicular Technology Conference*, vol. 2, pp. 691–695, May.

[25] Weichselberger, W., Herdin, M. Özcelik, H. and Bonek, E. (2006) A stochastic MIMO channel model with joint correlation of both link ends. *IEEE Transactions on Wireless Communications*, **5**(1), 90–101..

[26] Kermoal, J.P, Schumacher, L., Pedersen, K.I, Mogensen, P.E and Frederiksen, F. (2002) A stochastic MIMO radio channel model with experimental validation. *IEEE Journal on Selected Areas in Communications*, **20**(6), 1211–1226.

[27] Molisch, A.F et al. (2008) A comprehensive standardized model for ultrawideband propagation channels. *IEEE Transactions on Antennas and Propagation*, **5**(51), 3151–3166.

[28] Chong, C.-C., Kim, Y.E, Yong, S.K and Lee, S.S (2005) Statistical characterization of the UWB propagation channel in indoor residential environment. *Wireless Communications and Mobile Computing*, **5**(5), 503–512 (special issue on UWB communications).

[29] Xu, H., Kukshya, V. and Rappaport, T.S (2002) Spatial and temporal characteristics of 60 GHz indoor channels. *IEEE Journal on Selected Areas in Communications*, **20**(3), 620–630.

[30] Anderson, C.R and Rappaport, T.S (2004) In-building wideband partition loss measurements at 2.5 and 60 GHz. *IEEE Transactions on Wireless Communications*, **3**(3), 922–928.

[31] Matic, D., Harada, H. and Prasad, R. (1998) Indoor and outdoor frequency measurements for MM-waves in the range of 60 GHz. *Proceedings of the IEEE Vehicular Technology Conference*, vol. 1, pp. 567–571, May.

[32] Moriatis N. and Constantinou, P. (2004) Indoor channel measurements and characterization at 60 GHz for wireless local area network applications. *IEEE Transactions on Antennas and Propagation*, **52**(12), 3180–3189.

[33] Fiacco, M., Parks, M., Radi, H. and Saunders, S.R (1998) Final Report: Indoor propagation factors at 17 and 60 GHz. Technical report and study carried out on behalf of the Radiocommunications Agency, University of Surrey, August.

[34] Radi, H., Fiacco, M., Parks, M.A N. and Saunders, S.R (1998) Simultaneous indoor propagation measurements at 17 and 60 GHz for wireless local area networks. *Proceedings of the IEEE Vehicular Technology Conference*, pp. 510–514, May.

[35] Yang, H., Smulders, P.F M. and Herben, M.H A.J (2007) Channel characteristics and transmission performance for various channel configurations at 60 GHz. *EURASIP Journal on Wireless Communications and Networking*, article ID 19613.

[36] Geng, S., Kivinen, J., Zhao, X. and Vainikainen, P. (2009) Millimeter-wave propagation channel characterization for short-range wireless communication. *IEEE Transactions on Vehicular Technology*, 58(1), 3–13.

[37] Kalivas, G., El-Tanany, M. and Mahmoud, S. (1995) Millimeter-wave channel measurements with space diversity for indoor wireless communications. *IEEE Transactions on Vehicular Technology*, 44, 494–505.

[38] Kajiwara, A. (1997) Millimeter wave indoor radio channel artificial reflector. *IEEE Transactions on Vehicular Technology*, 46, 486–493.

[39] Bultitude, R.J C., Hahn, R.F and Davies, R.J (1998) Propagation considerations for the design of the an indoor broad band communications system at EHF. *IEEE Transactions on Vehicular Technology*, 47, 235–245.

[40] Kivinen, J. (2007) 60-GHz wideband radio channel sounder. *IEEE Transactions on Instrumentation and Measurement*, **56**(5), 1831–1838.

[41] Pendergrass, M. (2002) Empirically based statistical ultra-wideband channel model. IEEE P802.15-02/240-SG3a.

[42] Bohdanowicz, A. (2000) Wideband indoor and outdoor radio channel measurements at 17 GHz. UBICOM Technical Report, January.

[43] Thomas, H.J et al. (1994) An experimental study of the propagation of 55 GHz millimeter waves in an urban mobile radio environment. *IEEE Transactions on Vehicular Technology*, **43**(1), 140–146.

[44] Yang, H.B, Herben, M.H A.J and Smulders, P.F M. (2005) Impact of antenna pattern and reflective environment on 60 GHz indoor radio channel characteristics. *IEEE Antennas and Wireless Propagation Letters*, **4**.

[45] Collonge, S., Zaharia, G. and Zein, G.E (2004) Influence of the human activity on wideband characteristics of the 60 GHz indoor radio channel. *IEEE Transactions on Wireless Communications*, **3**(6), 2389–2406.

[46] Kunisch, J., Zollinger, E. Pamp, J. and Winkelmann, A. (1997) MEDIAN 60 GHz wideband indoor radio channel measurements and model. *Proceedings of the IEEE Vehicular Technology Conference*, pp. 2393–2397.

[47] Maltsev, A. et al. (2010) Channel models for 60 GHz WLAN systems. IEEE 802.11-09-0334-06-ad, January.

[48] Zwick, T., Beukema T.J and Nam, H. (2005) Wideband channel sounder with measurements and model for the 60 GHz indoor radio channel. *IEEE Transactions on Vehicular Technology*, **54**, 1266–1277.

[49] Choi, M.S, Grosskopf, G. and Rohde, D. (2005) Statistical characteristics of 60 GHz wideband indoor propagation channel. *Proceedings of the IEEE International Symposium on Personal, Indoor and Mobile Radio Commununications*, September.

[50] Saleh, A. and Valenzuela, R. (1987) A statistical model for indoor multipath propagation. *IEEE Journal on Selected Areas in Communications*, **5**(2), 128–137.

[51] Spencer, Q., Jeffs, B.D Jensen, M.A and Swindlehurst, A.L (2000) Modeling the statistical time and angle of arrival characteristics. *IEEE Journal on Selected Areas in Communications*, **18**(3), 347–360.

[52] Chong, C.-C., Tan, C.-M., Laurenson, D.I, McLaughlin, S., Beach, M.A and Nix, A.R (2003) A new statistical wideband spatio-temporal channel model for 5- GHz band WLAN systems. *IEEE Journal on Selected Areas in Communications*, **21**(2), 139–150.

[53] Chong, C.-C. and Yong, S.K (2005) A generic statistical-based UWB channel model for high-rise apartments, *IEEE Transactions on Antennas Propag.*, **53**(8), 2389–2399.

[54] Sawada, H. et al. (2006) LOS office channel model based on TSV model. IEEE 802.15-06-0377-00-00-3c, September.

[55] Pollock, T. et al. (2006) Office 60 GHz channel measurements and model. IEEE 802.15-06-0316-00-00-3c, July.

[56] Liu, C. et al. (2006) NICTA indoor 60 GHz channel measurements and analysis update. IEEE 802.15-06-0222-00-00-3c, May.

[57] Cramer, R.J, Scholtz, R.A and Win, M.Z (2002) An evaluation of the ultra-wideband propagation channel. *IEEE Transactions on Antennas and Propagation*, **50**(5), 561–570.

[58] Guerin, S. (1996) Indoor wideband and narrowband propagation measurements around 60.5 GHz in an empty and furnished room. *IEEE Vehicular Technology Conference '96*, vol. 1, pp. 160–164, May.

[59] Purwaha, J., Mank, A., Matic, D., Witrisal, K. and Prasad, R. (1998) Wide-band channel measurements at 60 GHz in indoor environments. *Symposium on Vehicular Technology and Communications*, Brussels, October.

[60] Park, J.H, Kim, Y., Hur, Y.S, Lim, K. and Kim, K.H (1998) Analysis of 60 GHz band indoor wireless channels with channel configurations. *IEEE International Symposium on Personal, Indoor and Mobile Radio Communications*, pp. 617–620.

[61] Smulders, P.F M. (1995) Broadband wireless LANs: a feasibility study. Ph.D. thesis, Eindhoven University.

[62] Manabe, T., Sato, K., Masuzawa, H., Taira, K., Ihara, T., Kasashima, Y. and Yamaki, K. (1995) Polarization dependence of multipath propagation and high-speed transmission characteristics of indoor millimeter-wave channel at 60 GHz. *IEEE Transactions on Vehicular Technology*, **44**(2), 268–274.

[63] Manabe, T., Miura, Y. and Ihara, T. (1996) Effects of antenna directivity and polarization on indoor multipath propagation characteristics at 60 GHz. *IEEE Journal on Selected Areas in Communications*, **14**(3), 441–448.

[64] Clavier, L. et al. (2001) Wideband 60 GHz indoor channel: characterization and statistical modeling. *Proceedings of the IEEE Vehicular Technology Conference*, vol. 4, pp. 2098–2102, October.

[65] Siamarou, A.G and Al-Nuaimi, M.O (2001) Multipath delay spread and signal level measurements for indoor wireless radio channel at 62.4 GHz. *IEEE Vehicular Technology Conference '01*, pp. 454–458.

[66] Hubner, J., Zerisberg, S., Koora, K., Borowski, J. and Finger, A. (1997) Simple channel model for 60 GHz indoor wireless LAN design based on complex wideband measurements. *IEEE Vehicular Technology Conference '97*, vol. 2, pp. 1004–1008, May.

[67] Davies, R., Bensebti, M., Beach,, M.A and McGeehan, J.P (1991) Wireless propagation measurements in indoor multipath environments at 1.7 GHz and 60 GHz for small cell systems. *IEEE Vehicular Technology Conference '91*, pp. 581–593, May.

[68] Williamson, M.R, Athanasiadou, G.E and Nix, A.R (1997) Investigating the effects of antenna directivity on wireless indoor communication at 60 GHz. *Proceedings of the IEEE Symposium on Personal, Indoor and Mobile Radio Communications*, vol. 2, pp. 635–639, September.

[69] Ranvier, S., Kivinen, J. and Vainikainen, P. (2007) Millimeter-wave MIMO radio channel sounder. *IEEE Transactions on Instrumentation and Measurement*, **56**(3), 1018–1024.

[70] Liu, C., Skafidas, E., Pollock, T.S and Evans, R.J (2006) Angle of arrival extended S-V model for the 60 GHz wireless desktop channel. *Proceedings of the IEEE International Symposium on Personal, Indoor and Mobile Radio Communications*, pp. 1–6, September.

[71] Moraitis, N. and Constantinou, P. (2006) Measurements and characterization of wideband indoor radio channel at 60 GHz. *IEEE Transactions on Wireless Communications*, **5**(4), 880–889.

[72] Molisch, A.F, Win, M.Z and Cassioli, D. (2000) The ultra-wide bandwidth indoor channel: from statistical model to simulations. *IEEE Journal on Selected Areas in Communications*, **20**, 1247–1257.

[73] Yong, S.K and Clerckx, B. (2007) The use of polarization for file transferring. ECMA TG20-TC32-2007-036, March.

[74] Oestges, C. and Clerckx, B. (2007) *MIMO Wireless Communications: From Real-World Propagation to Space-Time Code Design*. Boston: Elsevier.

[75] Maltsev, A. et al. (2009) Polarization model for 60 GHz. IEEE 802.11-09-0431-00ad, April.

[76] Pagani, P., Siaud, I., Malhouroux, N. and Li, W. (2006) Adaptation of the France Telecom 60 GHz channel model to the TG3c framework. IEEE 802.15-06-0218-00-003c, April.

[77] Sawada, H., Kato, S. and Sato, K. (2009) Propagation measurements and considerations for TGad channel modeling in conference room, living room and cubicle environments.IEEE 802.11-09-0874-01-ad, May.

[78] Toyoda, I. (2006) Reference antenna model with sidelobe level for TG3c Evaluation. IEEE 802.15.06-0474-00-003c, October.

3

Non-Ideal Radio Frequency Front-End Models in 60 GHz Systems

Chang-Soon Choi, Maxim Piz and Eckhard Grass

As described in Chapter 1, 60 GHz wireless systems are now seeing various kinds of consumer applications from short-range file transfer to indoor Gbps home networks.[1] One of the key technical challenges for the widespread commercial use of 60 GHz based wireless end-user products is their cost-effective implementation. This has provided the motivation for a great deal of work on silicon-based millimeter-wave technologies capable of providing higher integration and lower power consumption than the III-V compound semiconductor technologies used in 60 GHz bands even a few years ago.

Nevertheless, there is still much to do if we are to make silicon-based 60 GHz front-ends competitive with microwave front-ends below 5 GHz. The challenging 60 GHz radio frequency (RF) front-ends induce several critical RF non-idealities that must be addressed in 60 GHz system design. This problem becomes even more critical in mobile or handheld device related applications in which an aggressive RF circuit design for low power consumption and small form factors is a must.

The aim of this chapter is to address RF non-idealities that should be considered in the 60 GHz system design and to establish their behavioral models, providing

[1]This work was funded by the German Federal Ministry of Education and Research (BMBF) through WIGWAM and EASY-A projects. The authors would like to thank the analog circuit team at IHP.

simple and efficient methods to predict system performance without computationally complex circuit level simulations. Section 3.1 gives an overview of main RF analog front-end architectures and discusses their applicability for 60 GHz systems. RF non-idealities in these architectures are also briefly introduced. Section 3.2 discusses the nonlinearity of power amplifiers. A brief review of power amplifier models is first given, and then their influences on system performance are presented. Section 3.3 deals with phase noise arising from local oscillators (LOs) and the behavioral models we use to characterize it. The chapter ends with a discussion of other RF impairments that may also affect system performance.

3.1   RF Front-End Architecture

3.1.1   Super-Heterodyne Architecture

The super-heterodyne architecture has long been a standard choice for wireless receivers due to its capability to provide high selectivity and sensitivity. Basically, it converts received RF signal energy into baseband through several steps, which makes it possible to optimize system parameters at each step. Figure 3.1(a) shows a block diagram of a two-step super-heterodyne receiver. An RF bandpass filter first rejects out-of-band signals among the RF signals collected by the antenna. The filtered in-band signals are then amplified by a low-noise amplifier (LNA) while keeping the system noise floor as low as possible. The next filter suppresses image signals which could be overlapped with original signals after mixing. The frequency down-conversion with the first LO moves RF signals down to the intermediate frequency (IF) band. Channel selection is done at the IF stage with the help of highly selective filters, which lessen the dynamic range requirement in the following stages. An automatic gain control (AGC) amplifier adjusts signal levels for optimum I/Q demodulator operation. After splitting into two signals, the IF signals are down-converted into baseband with two mixers and the second quadrature LOs. After additional gain-control and anti-aliasing filtering, the baseband signals are converted to digital.

   In addition to high channel selectivity, this results in much less LO self-mixing which deteriorates receiver dynamic range, because the first LO is different from RF. There might still be some LO leakage of the second LO to the IF I/Q mixer, but we can expect good isolation in the low-IF region and there is no high gain component in IF stages. Another benefit would be the possibility of a fixed frequency LO at the first LO stage (not all super-heterodyne). It does not require a voltage-controlled oscillator (VCO) with a wide tuning range and its phase-locked loop (PLL) can be optimized to suppress the phase-noise contribution from the VCO, which gives improved phase-noise characteristics than a tunable frequency synthesizer covering all the RF channels.

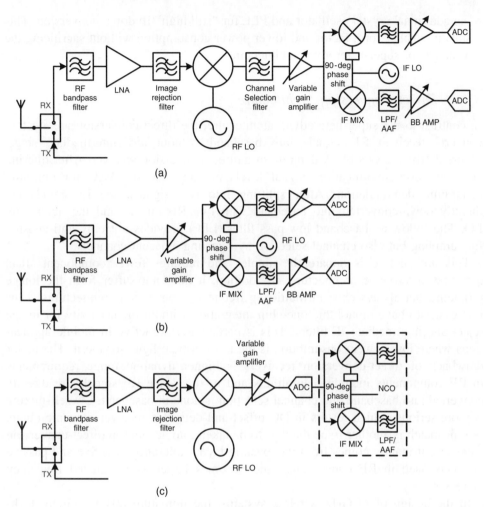

Figure 3.1 RF analog front-end: (a) super-heterodyne; (b) direct conversion; and (c) low-IF receiver architecture.

Due to its superior performance, most 60 GHz transceivers developed so far are based on the super-heterodyne architecture. Image rejection problems can be alleviated by incorporating high IF of more than a few gigahertz with the frequency selective responses of an LNA or power amplifier (PA) without integrating an external image rejection filter. For example, there has been work reporting 5, 9 and 12 GHz IF for 60 GHz transceivers [1–3]. The use of 5 GHz has proved the most popular of these since the IF building blocks could be directly used for 5 GHz wireless LAN applications. Alternatively, this IF can be generated from the first LO by frequency division – the sliding IF architecture. This potentially eliminates the use of an additional crystal oscillator and PLL for "still high" IF down-conversion. This results in smaller form factor and lower power consumption without sacrificing the advantages of super-heterodyne architecture.

3.1.2   Direct-Conversion Architecture

In contrast to the super-heterodyne architecture, the direct-conversion architecture converts received RF signals into baseband without any intermediate stage. Figure 3.1(b) shows a block diagram of a direct-conversion receiver. After filtering and low-noise amplification, signal levels are adjusted by AGC for optimum quadrature demodulation. After splitting into two signals, the IF signals are directly down-converted into baseband with two RF mixers and the quadrature LO. The following baseband low-pass filter (LPF) provides not only anti-aliasing for sampling but also channel selection by rejecting adjacent channels.

This architecture is apparently simpler and requires fewer components than the super-heterodyne architecture. In addition, it does not suffer from the image problems that always exist in multi-step frequency conversion architectures. Thus it is easier to have monolithic one-chip integration without the necessity for image reject and intermediate IF filters. This is particularly attractive in mobile applications where low power consumption is the most critical figure-of-merit. The major drawback of direct-conversion receivers is higher dynamic range requirements in RF components including amplifiers and mixers since RF signals are directly converted into baseband and channel selection is done in baseband. LO self-mixing is more serious, which results in DC offset and degraded receiver dynamic range. I/Q mismatch is another issue that has to be taken into account in direct-conversion receivers. It mainly comes from the existing layout tolerance of active and passive elements used in RF components, and becomes larger as designated frequency increases.

In the design of 60 GHz wireless systems, the immature 60 GHz circuit technologies have made it more challenging to use the direct-conversion architecture. Nevertheless, several results have been reported especially for short-range file transfer applications that require low-cost transceivers with low power consumption [4–5].

3.1.3   Low-IF Architecture

The low-IF architecture is also called "digital IF" since down-converted IF signals are digitally processed after analog-to-digital conversion. A block diagram is shown in Figure 3.1(c). RF signals are not directly converted into baseband but moved to the low-IF band where the analog-to-digital converter (ADC) sampling rate can support the conversion into digital signals. Then, these low-IF signals are sampled with ADC and I/Q quadrature demodulation is done in the digital domain. This was originally proposed to overcome the problems of the direct-conversion architecture while keeping the simple RF architecture. However, this increases the burden of ADC which requires both higher sampling rate and higher dynamic range.

The ADC sampling rate has already been a bottleneck for Gbps 60 GHz wireless systems. For example, the IEEE 802.15.3c standard requires a sampling rate in excess of 2.6 GHz [6]. However, there are few ADCs available for this sampling rate which satisfy the required number of bits. Low IF would need more than 5 GHz sampling with more than 8 bits, which is hardly realizable with current technologies. Thus, low-IF architectures are currently not in the mainstream of 60 GHz system design.

3.2 Nonlinear Power Amplifier

3.2.1 Tradeoff Between Linearity and Efficiency

Most active RF components in transmitters and receivers exhibit nonlinear transfer characteristics. Among them, the power amplifier is known as the most dominant nonlinear source in a wireless transmitter. This problem is more serious in the 60 GHz wireless systems of interest in this book since silicon-based integrated power amplifiers which are desirable for low-cost solutions cannot provide enough radiation power transmission while keeping high linearity.

Figure 3.2 shows one example of a 60 GHz power amplifier fabricated by the SiGe process, which plots output power and power added efficiency (PAE) as a

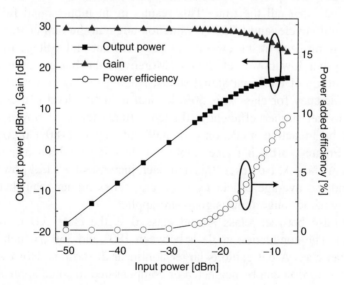

Figure 3.2 Transfer curve of a 60 GHz SiGe power amplifier.

function of input power [7]. The PAE, denoted by η, is defined as the ratio of amplified output signal power to DC power applied to the power amplifier, which can be expressed as

$$\eta = \frac{P_{out} - P_{in}}{P_{dc}}, \qquad (3.1)$$

where P_{in} is the input power, P_{out} is the desired output power of a power amplifier in the band of interest, and P_{dc} is the DC input power supplied to the power amplifier.

In modern wireless communication systems, the power amplifier accounts for the largest portion of system power consumption, thus it is highly favorable to drive a power amplifier close to output saturation point in order to get maximum power efficiency, shown on the right-hand axis of Figure 3.2. However, this increases the nonlinearity of the power amplifier, resulting in increased distortion and degraded signal-to-noise-and-distortion ratio (SNDR). Apparently, there is a tradeoff between power efficiency and nonlinearity imposed on the power amplifier. This is particularly important for a system designer to take into account in deciding the optimum driving level of a power amplifier that fulfills system requirements.

This tradeoff also depends on the class of the power amplifier. Classes A, B, AB and C are the most widely used power amplifiers. The classification is determined by input signal level and the operation point of an amplifier. Simplified illustrations of the operation conditions of these power amplifiers are given in Figure 3.3.

In class A power amplifiers, as shown in Figure 3.3(a), the core transistor is biased to be in active mode all the time. This results in an input signal fully amplified through its entire cycle, and a $360°$ conduction angle at the output stage. Nonlinear distortion is minimized at the expense of large current and voltage bias, implying high linearity but low power efficiency. Moreover, the biasing current is always applied regardless of the input signal status, leading to low power efficiency. The maximum efficiency for class A is 50% if ideal inductive load is used.

In order to get higher efficiency, the core transistor in a power amplifier is designed to be in active mode only in half duty cycle ($180°$) of input signals. As seen in Figure 3.3(b), it is inevitable for class B to exhibit a higher level of distortion than class A. However, this approach increases the ideal power efficiency to 78.5% and the average efficiency is several times higher than that of class A when high dynamic range input signals are applied.

A compromise between class A and class B is the class AB mode shown in Figure 3.3(c). The conduction angle is between $180°$ and $360°$, which gives better efficiency than class A but exhibits larger nonlinear distortions. However, the level of nonlinear distortion can be neglected or compensated in some applications, particularly in systems employing a single carrier (SC) modulation scheme. The bias

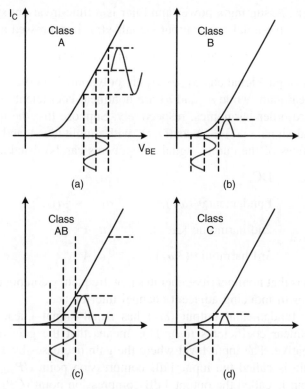

Figure 3.3 Classes of power amplifiers: (a) class A; (b) class B; (c) class AB; (d) class C.

level is just above the threshold level and can be adjusted according to linearity requirements.

Class C is a more power-efficient solution than class B. The transistor is biased below the threshold voltage (in cutoff mode) so that it becomes active only for less than half the cycle of input signals as illustrated in Figure 3.3(d). Theoretically, the power efficiency can be up to 100% by simply decreasing the conduction angle to 0°. However, this significantly decreases the output power level. Generally, class C modes can achieve a power efficiency of 90% by using a 90° conduction angle.

3.2.2 Nonlinearity Modeling

In order to quantify the influence of PA nonlinearity on system performance, a behavioral model for PA characteristics is required. In an RF front-end specification, gain compression and inter-modulation distortion levels are normally described to express the level of nonlinearity. These system parameters can be derived by simple

analytical models. Assuming a power amplifier is a time-invariant system modeled
with third-order polynomials, the output signal, $y(t)$, is expressed as

$$y(t) = a_1 x(t) + a_2 x^2(t) + a_3 x^3(t), \tag{3.2}$$

where $x(t)$ is the input signal and a_1, a_2, a_3 are polynomial coefficients: a_1 indicates
small signal linear gain, while a_2 and a_3 are nonlinear coefficients attributed to the
second- and third-order harmonics, respectively. Suppose that the input signal is a
single sinusoidal signal, $x(t) = A \cos(\omega_f t)$, with amplitude A and frequency ω_f;
then the amplitudes of the output signal components can be listed as:

$$
\begin{aligned}
&\text{DC,} && \tfrac{1}{2} a_2 A^2; \\
&\text{Fundamental } (\omega_f), && a_1 A + \tfrac{3}{4} a_3 A^3; \\
&\text{2nd harmonic } (2\omega_f), && \tfrac{1}{2} a_2 A^2; \\
&\text{3rd harmonic } (3\omega_f), && \tfrac{1}{4} a_3 A^2.
\end{aligned}
\tag{3.3}
$$

The implication is that nonlinearity generates new frequency components that appear
as spectral regrowth inducing adjacent channel interference.

The gain for fundamental output (ω_f) has small signal linear gain (a_1) and
third-order nonlinear coefficients (a_3). This means that the gain decreases when
a_3 becomes negative. The input level where the gain decreases by 1 dB from small
signal linear gain is called the input 1 dB compression point $(IP_{1\,dB})$ and the corresponding output is called the output 1 dB compression point $(OP_{1\,dB})$. The $IP_{1\,dB}$
is given by

$$IP_{1\,dB} = \sqrt{0.145 \left| \frac{a_1}{a_3} \right|}. \tag{3.4}$$

This nonlinearity becomes more critical when multi-tone input signals are applied
since it gives rise to frequency mixing components between multi-tone signals,
typically called inter-modulation distortion (IMD). Assuming two-tone signals are
applied, i.e. $x(t) = A \cos(\omega_1 t) + A \cos(\omega_2 t)$, the amplitudes of fundamental and
inter-modulation components can be expressed as follows: fundamental (ω_1),

$$(a_1 A + (9a_3/4)A^3) \cos(\omega_1 t), \quad (a_1 A + (9a_3/4)A^3) \cos(\omega_2 t);$$

second-order inter-modulation,

$$a_2 A^2 \cos((\omega_1 + \omega_2)t) + a_2 A^2 \cos((\omega_1 - \omega_2)t);$$

third-order inter-modulation,

$$(3a_3/4)A^3 \cos((2\omega_1 + \omega_2)t) + (3a_3/4)A^3 \cos((2\omega_1 - \omega_2)t)$$
$$+ (3a_3/4)A^3 \cos((\omega_1 + 2\omega_2)t) + (3a_3/4)A^3 \cos((\omega_1 - 2\omega_2)t)$$

Here, the second-order inter-modulation can be removed simply by employing a fully differential design of circuits or bandpass filter because it is far from the band of interest. On the other hand, the third-order inter-modulation components at $(2\omega_1 - \omega_2)$ and $(2\omega_2 - \omega_1)$ lie in the band of interest and cannot be eliminated by applying a filter. These components degrade the quality of signals like a noise. The spurious free dynamic range (SFDR) is commonly used to specify an input power level which is only dominated by thermal noise level, free of such distortion. It also introduces the third-order intercept point (IP3), where the third-order inter-modulation components have the same power as the fundamental components. Note that this does not exist in practice because a power amplifier becomes saturated before reaching that point. The IP3 is obtained by linear interpolation of small signal outputs for fundamental and third-order inter-modulation components. A graphical view of these parameters is shown in Figure 3.4.

3.2.3 Behavioral Models

In the previous section, we only investigated how the input signal level (amplitude) influences the output signal amplitude in a nonlinear power amplifier. However, the

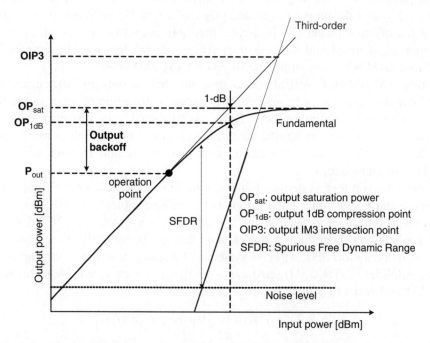

Figure 3.4 Third-order interception point (IP3), 1 dB compression point and spurious free dynamic range measurements in the transfer curve of active components.

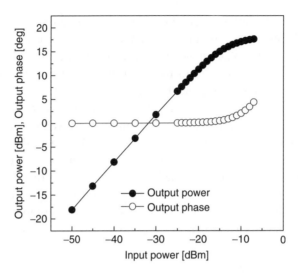

Figure 3.5 Output power and output phase as a function of input power in a 60 GHz SiGe power amplifier.

output phase of a power amplifier also changes as a function of input signal level. Figure 3.5 shows the output power and phase of a 60 GHz SiGe power amplifier as a function of input power [7]. It is clear that both output power and phase depend on input signal level and these characteristics can be described by amplitude to amplitude (AM/AM) and amplitude to phase (AM/PM) effects.

If these AM/AM and AM/PM effects are connected not only with the current input signal but also with the previous one, the nonlinear distortion of a power amplifier is also affected by memory effects. These are attributed to high temperature, rapid thermal variation and biasing fluctuation, mainly occurring in high-power (several-watt-level) amplifiers used in satellite communications and in base stations for cellular communications.

In 60 GHz wireless systems of interest in this book, memory effects can be neglected since transmitter emission power is limited typically to 10 dBm in most countries as described in Section 1.4 to support indoor wireless LAN and wireless PAN applications. Therefore, a power amplifier can be modeled as a memoryless system in indoor 60 GHz wireless systems. Assuming that the input signal to a power amplifier is $x(t) = A(t) \cos[\omega_f t + \theta(t)]$, the output signals characterized by AM/AM and AM/PM effects can be expressed as

$$y(t) = G[A(t)] \cos \left[\omega_f t + \theta(t) + \phi[A(t)] \right], \tag{3.5}$$

where $G[A(t)]$ and $\varphi[A(t)]$ are the AM/AM and AM/PM effects of a power amplifier, respectively.

Several behavioral models have been reported in the literature to describe the nonlinearity of power amplifiers. The Saleh model was originally proposed for traveling-wave tube (TWT) amplifiers commonly used in satellite communications [8]. It is composed of two equations describing AM/AM and AM/PM effects, expressed respectively as

$$G[A(t)] = \frac{\alpha_G |A(t)|}{1 + \beta_G |A(t)|^2} \tag{3.6}$$

and

$$\varphi[A(t)] = \frac{\alpha_\varphi |A(t)|^2}{1 + \beta_\varphi |A(t)|^2}, \tag{3.7}$$

where α_G, β_G, α_φ, and β_φ are real number coefficients. The Saleh model has been widely used in the design of power amplifier linearization techniques and reference models for proposal evaluation in IEEE standards societies. However, it does not fit well with solid-state power amplifiers such as GaAs FET and SiGe HBT amplifiers of interest in 60 GHz wireless systems because this model becomes inaccurate in the saturation regions of such amplifiers. Figure 3.6 shows the comparison between the measurement data and the fitted Saleh model for 60 GHz SiGe amplifiers. While the Saleh model fits very well at low input voltages, the deviation becomes significant at higher input voltages that are close to the saturation of the amplifier.

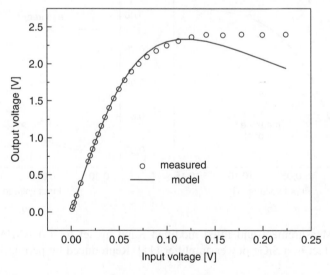

Figure 3.6 Measured output voltage of a 60 GHz SiGe power amplifier and fitted Saleh model.

Another model, the Rapp model, is better suited to solid-state power amplifiers [9]. It gives a simple and intuitive equation made up of amplifier small signal gain, G, saturation voltage, V_{sat} and a smoothing factor, p, which is given by

$$G[A(t)] = \frac{G|A(t)|}{(1 + (G|A(t)|/V_{sat})^{2p})^{1/2p}} \tag{3.8}$$

for the AM/AM effect. The Rapp model only considers the AM/AM effect and assumes that the AM/PM effect is negligible in solid-state power amplifiers. However, in some cases, the AM/PM effect cannot be completely neglected and should be included to accurately evaluate system performance. In order to take the AM/PM effect into account, a modified Rapp model was developed in [10]. The AM/AM effect of the modified Rapp model is the same as in Equation (3.8) while the AM/PM effect is given by

$$\phi[A(t)] = \frac{\alpha|A(t)|^q}{1 + (|A(t)|/\beta)^q}, \tag{3.9}$$

where α, β and q are fitting parameters. Figure 3.7 shows the measurement results and the fitted modified Rapp model for a 60 GHz SiGe power amplifier. It can clearly be seen that the modified Rapp model fits the measured data well and thus accurately characterizes the nonlinear behaviors of 60 GHz power amplifiers.

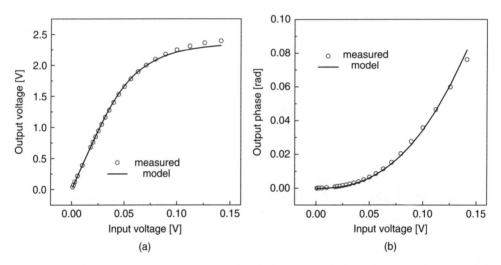

Figure 3.7 Measurement and fitted modified Rapp model for (a) AM/AM effect and (b) AM/PM effect in a SiGe power amplifier [11]. Reproduced by permission of © 2009 IEEE.

3.2.4 Output Backoff Versus Peak-to-Average Power Ratio

As seen in Figure 3.5, nonlinear distortion strongly depends on the operation point of a power amplifier. By decreasing the output power of a given power amplifier, the power amplifier can deliver more linear output at the expense of less power efficiency. Within the range of operation points where linearity given by a power amplifier meets the requirements, it is maximized for higher power efficiency directly, resulting in low power consumption. To characterize this, output backoff (OBO) from the output saturation point is usually introduced, which can be expressed as

$$OBO = 10 \log_{10}(P_{out}/P_{sat}) \qquad (3.10)$$

where P_{out} is the average output power emitted from a power amplifier and P_{sat} is the saturation output power. The OBO is used to quantify how much output power a power amplifier provides in comparison to maximum achievable power.

The requirements for OBO are mainly determined by input signals that pass through an amplifier. The peak-to-average power ratio (PAPR), which is defined as the power ratio of peak power to average power in the time domain, is commonly used to characterize the dynamic range of input signals. Figure 3.8(a) gives one

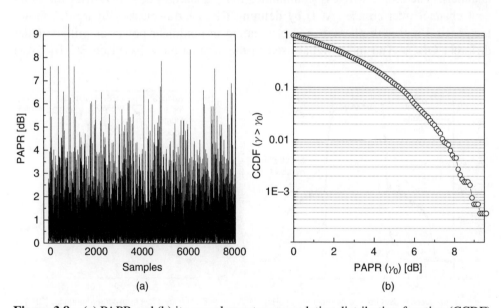

(a) (b)

Figure 3.8 (a) PAPR and (b) its complementary cumulative distribution function (CCDF) for OFDM signals.

example of the PAPR for Gbps orthogonal frequency division multiplexing (OFDM) signals in 60 GHz wireless systems. It can be seen that the PAPR significantly varies with input signal samples. Figure 3.8(b) shows the complementary cumulative distribution function (CCDF) of an OFDM signal PAPR. The CCDF of the PAPR is defined as the probability that a given PAPR is larger than a threshold PAPR value in the time domain.

Larger OBO is indispensable for higher PAPR input signals. However, this imposes lower power efficiency and increases implementation costs. In addition, larger OBO also increases the requirements of data converters, so that large peak signals can be precisely interpreted between analog and digital domains.

3.2.5 Impact of Nonlinear Power Amplifier

For a given input signal with a specific PAPR value, the OBO of a power amplifier is adjusted to meet the RF spectral mask and error vector magnitude (EVM) typically defined in many wireless communication standards. The third-order and higher odd-order inter-modulation induces not only nonlinear distortions in the band of interest but also causes spectral regrowth in adjacent channels. This out-of-band nonlinear distortion may seriously interfere with other signals in the adjacent channels, Therefore, wireless communication standards need to restrict the adjacent channel interference (ACI) by defining RF spectral masks. Figure 3.9 shows the spectral regrowth phenomenon originating in nonlinear power amplifiers under different OBOs. The input signal used in the simulation is based on SC 16-QAM

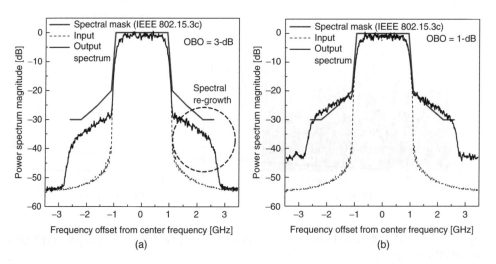

Figure 3.9 Spectral regrowth and its induced adjacent channel interference under (a) 3 dB and (b) 1 dB output backoff.

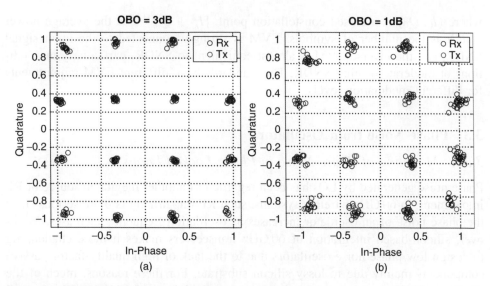

Figure 3.10 Constellations of 16-QAM modulated signals before and after nonlinear amplification under (a) 3 dB and (b) 1 dB output backoff.

modulation and the power amplifier used is a 60 GHz SiGe power amplifier as shown in Figure 3.7. The spectrum of the input signal and the spectral mask defined in the IEEE 802.15.3c standard are included in the same figures for comparison [6]. It can clearly be seen that the spectral regrowth in out-of-band regions is below the spectrum mask when the OBO is 3 dB. However, an OBO of 1 dB does not fulfill the spectral mask requirement defined in IEEE 802.15.3c and thus has to be increased.

Furthermore, the nonlinear distortion in the band of interest also results in input signal degradation. The constellations of 16-QAM signals before and after nonlinear amplification under 3 dB and 1 dB OBO are show in Figure 3.10. Higher distortion is observed in output constellation when less OBO is applied in Figure 3.10(b). It can also be seen that the outer points appear more or less in a large ring shape, which arises from the gain saturation effects of a power amplifier. In addition, the constellations are slightly rotated in the clockwise direction, particularly under the condition of low OBO (1 dB) as a result of the AM/PM effect observed in Figure 3.5.

The accuracy of the modulated signal after passing through a nonlinear power amplifier can be characterized by EVM measurement, widely used in IEEE 802.11 and IEEE 802.15. The EVM is defined as the magnitude of the error vector with respect to the input signal vector,

$$EVM = \frac{\sum \left[(I - I_0)^2 + (Q - Q_0)^2 \right]^{1/2}}{N \cdot [I_m^2 + Q_m^2]^{1/2}}, \tag{3.11}$$

where (I_0, Q_0) is a closed constellation point, $[I_m^2 + Q_m^2]^{1/2}$ is the average power and N is the number of symbols. EVM is strongly connected to the input signal SNDR, and the inverse EVM is almost equivalent to SNDR, particularly in conditions of moderate or high SNDR. The standards have different EVM requirements for different modulation levels.

3.3 Phase Noise from Oscillators

3.3.1 Modeling of Phase Noise in Phase-Locked Loops

Phase noise generated by LOs has been regarded as another inevitable source of RF impairment. In 60 GHz wireless systems, such RF impairment is more serious since the phase noise of an oscillator increases as oscillation frequency increases. Moreover, silicon-based integration of 60 GHz transceivers makes it more challenging to design low phase-noise oscillators due to the lack of high quality-factor passive components mainly due to lossy silicon substrate. For these reasons, much of the work on 60 GHz has focused on the design methodology of 60 GHz PLLs as well as optimization of baseband schemes to make the system less susceptible to phase noise. In order to take into account phase noise in PLLs, the analysis and modeling of a generic PLL are presented in this section.

A simplified diagram of a PLL is shown in Figure 3.11. It consists of a reference crystal oscillator, a phase/frequency detector (PFD), a charge pump (CP), a low-pass filter (LPF), a VCO and a programmable divider. In a PLL, the VCO output phase is divided by N and compared with a reference phase in the PFD. A charge pump provides current proportional to the phase error. It is then filtered by an LPF and applied to the control port of a VCO. Suppose that the output voltage of a PLL is

$$V_{out}(t) = V_0 \cos[\omega_0 t + \phi_{out}(t)], \tag{3.12}$$

where V_0 is the amplitude, $f_0 = \omega_0/2\pi$ is the oscillation frequency, and $\phi_{out}(t)$ is the excess phase. All components in a PLL basically contribute to the overall

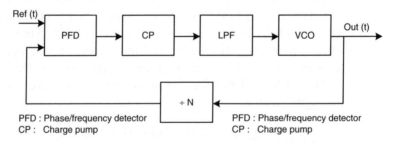

Figure 3.11 A generic phase-locked loop architecture.

system phase noise [12]; however, a reference clock (with buffer) and a VCO are the most dominant sources of PLL phase noise. The transfer characteristics of these phase-noise sources to output phase noise are investigated with the following system models [12, 13].

Figure 3.12 shows the linearized model of a PLL, where the $I_{CP}/2\pi$ [A/rad] is the PFD/CP gain with charge pump current I_{CP}, $F(s)$ is the loop filter transfer function, K_{VCO}/s is the VCO gain with the frequency factor of s, and N is the divider ratio of a PLL. The value $K_{VCO} = d\omega/dV_{ctrl}$ refers to the angular frequency in rad/s/V. With this PLL model, the closed-loop transfer function of a PLL can be described as

$$H = \frac{H_0}{1 + H_0/N}, \tag{3.13}$$

where the forward transfer function, H_0, is

$$H_0 = \frac{I_{CP}F(s)K_{VCO}}{2\pi s}. \tag{3.14}$$

The second-order low-pass filter considered in this simulation is shown in Figure 3.13(a) with capacitor values of C_1, C_2 and resistor value of R_1. It has trans-impedance characteristics of

$$F(s) = \left(R_1 + \frac{1}{sC_1}\right) \left\| \left(\frac{1}{sC_2}\right)\right. . \tag{3.15}$$

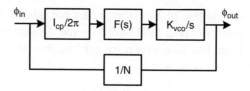

Figure 3.12   A linearized PLL model.

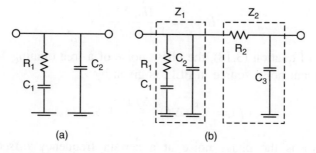

Figure 3.13   (a) Second-order and (b) third-order loop filter.

Most PLL architectures employ higher-order low pass filters in order to suppress reference spurs. Figure 3.13(b) gives one example of a third-order loop filter with capacitor values of C_1, C_2, C_3 and resistor values of R_1, R_2. It has transfer function

$$F(s) = \frac{1}{sC_3} \frac{1/Z_2}{1/Z_1 + 1/Z_2},$$ (3.16)

in which

$$Z_1 = \left(R_1 + \frac{1}{sC_1} \right) \left\| \left(\frac{1}{sC_2} \right), \quad Z_2 = \left(R_2 + \frac{1}{sC_3} \right).$$ (3.17)

The first dominant noise source is a reference oscillator. The reference oscillator phase noise as a function of frequency offset (f) can be modeled as

$$S_{ref}(f) = S_{ref}(\Delta f) \frac{(\Delta f)^2}{f^2} + S_{ref,floor},$$ (3.18)

where $S_{ref}(\Delta f)$ is the phase noise at a certain frequency offset, Δf, in the -20 dB/decade region of the spectrum and $S_{ref,floor}$ is the noise floor of the reference. Assuming no noise contribution from VCO in Figure 3.12, the transfer function from a reference phase to a PLL output phase is given by

$$\frac{\phi_{ref}^{out}}{\phi_{ref}} = \frac{H_0}{1 + H_0/N},$$ (3.19)

where ϕ_{ref}^{out} is the PLL output phase from reference phase ϕ_{ref}, H_0 is the forward transfer function described in (14) and N is the divider ratio. Equation (3.19) indicates a low-pass filter function given by a PLL, therefore the reference noise spectrum at PLL output is given by

$$S_{ref}^{out} = S_{ref} N^2 |H_{LPF}|^2,$$ (3.20)

in which

$$H_{LPF} = \frac{H_0/N}{1 + H_0/N}.$$ (3.21)

Analogous to Equation (3.18), the phase noise of a free-running VCO, which is the other dominant noise source in PLL, is given by

$$S_{vco}(f) = S_{vco}(\Delta f) \frac{(\Delta f)^2}{f^2} + S_{vco,floor},$$ (3.22)

where $S_{vco}(\Delta f)$ is the phase noise at a certain frequency offset, Δf, in the -20 dB/decade region of the spectrum (typically 1 MHz for VCO) and $S_{vco,floor}$ is the noise floor of the reference. Note that flicker noise is disregarded in Equation (3.22) because we only focus on applications in which the loop bandwidth is larger than the $1/f^3$ corner frequency of the VCO. Moreover, this flicker noise in the VCO is effectively suppressed by the PLL. The phase-noise contribution from the VCO to PLL output is clearly different from that of a reference clock. The transfer function from VCO output phase is given by

$$\frac{\phi_{vco}^{out}}{\phi_{vco}} = \frac{1}{1 + H_0/N}.$$ (3.23)

Thus, the VCO noise spectrum at the PLL output is given by

$$S_{vco}^{out} = S_{vco}|1 - H_{LPF}|^2,$$ (3.24)

where H_{LPF} is the low-pass filter function as described in Equation (3.21). In contrast to a reference clock noise, a VCO noise is high-pass filtered by a PLL.

The total phase-noise spectrum of a PLL is the summation of noise spectra from individual components since they are totally uncorrelated. Figure 3.14 shows the simulated phase-noise contributions from two major sources, reference oscillator and VCO, at a PLL output using the aforementioned analysis. The total phase noise at PLL output is also included for comparison purpose. It is clear that the reference oscillator is the dominant noise source at low-frequency offsets. As the frequency offset increases, the contribution from VCO becomes more dominant.

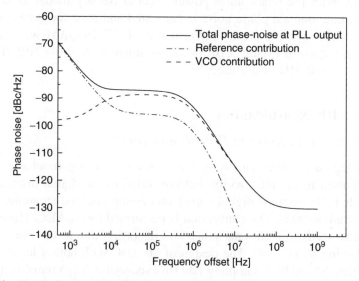

Figure 3.14 Total phase-noise spectrum at PLL output and its contributions from a reference clock and a VCO.

Note that this phase-noise spectral density can be minimized by optimum design of a PLL loop filter as shown in Equations (3.12)–(3.24). The optimum PLL bandwidth is a result of a tradeoff between the low-pass filtered reference noise and the VCO noise which is high-pass filtered by a PLL operation.

3.3.2   Behavioral Modeling for Phase Noise in Phase-Locked Loops

Given the phase-noise spectrum in Figure 3.14, the reference noise contribution at low frequency offset can be reasonably neglected for 60 GHz system simulation. The main reason for this is that the reference noise dominates low-frequency phase noise that does not affect system performance of high-rate SC systems, and even OFDM systems supported by common phase-error correction. In addition, the phase noise of reference is typically much lower than that of the VCO by many orders of magnitude. Consequently, the PLL output phase noise can be simply modeled as the high-pass filtered VCO phase noise [10], which yields a simple one-pole and one-zero behavioral model given by

$$PSD(f) = K \cdot \frac{[1 + (f/f_z)^2]}{[1 + (f/f_p)^2]},$$ (3.25)

where K is the low frequency phase noise below the loop-filter bandwidth of the PLL, f_p is the pole frequency (i.e. the cutoff frequency of the high-pass filter given by the PLL) and f_z is the zero frequency determined by the noise floor of the VCO. With the phase-noise power spectral density model as described in Figure 3.15, time-domain phase noise is generated and used for system simulation. The effect of phase noise on the performance of SC 16-QAM with symbol rate 1.728 GHz is shown in Figure 3.16. In this simulation, $K = -87$ dBc/Hz, $f_p = 1$ MHz and $f_z = 100$ MHz were used.

3.4   Other RF Non-Idealities

3.4.1   Quantization Noise in Data Converters

In modern digital communication systems, received analog baseband signals are not directly used to get information but converted to the digital domain and then processed digitally because digital signal processing confers economic as well as performance advantages. The conversion is performed by an ADC. However, ADC operation also leads to additional noise and distortion due to the finite number of bits and clipping in an ADC. The design of an ADC with both a larger number of bits (more than 8) and high sampling rate (in excess of 3 Gsps) remains a challenge. This is one of the serious bottlenecks for commercial use of multi-gigabit 60 GHz

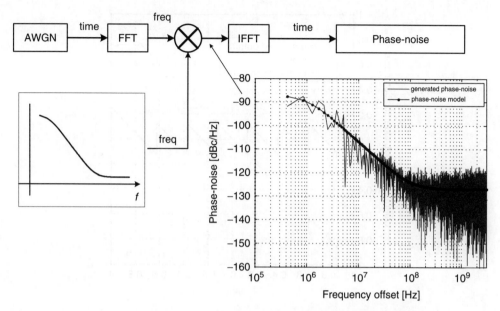

Figure 3.15 Time-domain phase-noise generation from the modelled phase-noise spectral density.

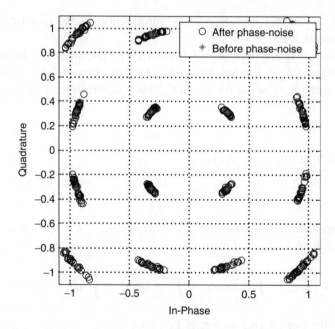

Figure 3.16 Constellations of 16-QAM modulated signals before and after suffering from phase noises.

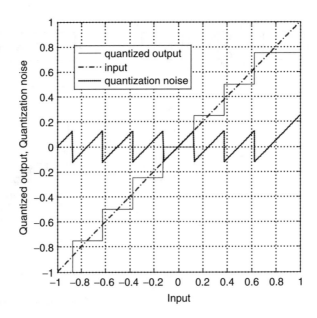

Figure 3.17 Quantized output and quantization noise of an analog-to-digital converter.

wireless systems. In this section a brief introduction to the quantization noise in ADCs is provided.

Suppose that N is the number of bits in the ADC and V_{ref} is the maximum reference voltage swing limited by ADC clipping. Figure 3.17 shows quantized output and quantization noise for an ADC. The quantization error is given by

$$e_q = \frac{\Delta}{2},\tag{3.26}$$

where the quantization level, Δ, is

$$\Delta = \frac{V_{ref}}{2^N}.\tag{3.27}$$

If an input signal is a zero-mean random variable, then average quantization noise power, e_{rms}^2, can be expressed as

$$e_{rms}^2 = \int_{-T/2}^{T/2} e^2\, dt = \int_{-T/2}^{T/2} \Delta^2 \left(\frac{-t}{T}\right)^2 dt = \frac{\Delta^2}{12}.\tag{3.28}$$

In this sense, the peak signal power is given by

$$P_{xPK} = 2^{2(N-1)}\Delta^2.\tag{3.29}$$

The average signal power is

$$P_{AVG} = \frac{P_{xPK}}{\eta} = \frac{2^{2(N-1)}\Delta^2}{\eta}, \tag{3.30}$$

where η is peak-to-average power ratio (PAPR) defined in the Section 3.2.4. This yields the average signal-to-quantization noise ratio (SQNR) of an ADC:

$$SQNR_{AVG} = \frac{P_{AVG}}{\sigma_e^2} = \frac{3(2^{2N})}{\eta} = 6.02N + 4.77 - 10\log_{10}(\eta). \tag{3.31}$$

If the input signal is ideal sinusoidal, giving a PAPR of 2, this equation becomes $6.02N + 1.76$[dB], which is known as the maximum SQNR of an ideal ADC. Higher PAPR results in lower SQNR, therefore it is necessary to have a larger number of bits in the case of a higher PAPR input signal in order to avoid SQNR degradation. This analysis also implies that 6 dB SNQR improvement is obtained by an additional bit of resolution in an ADC. The ADC requirements in terms of number of bits and clipping level are determined by SNR requirements given by the input signals. Figure 3.18 shows the output SNR after ADCs with different number of bits. The simulation was done with SC 16-QAM transmission through 60 GHz line-of-sight (LOS) office channel, CM3 (See Chapter 2). At low input SNR, the total output noise is largely dominated by channel noise and almost independent of the number of bits in an ADC. On the other hand, in the case of high input SNR, the output SNR is severely limited by quantization noise and improves with

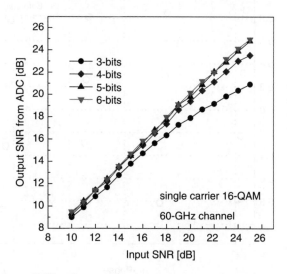

Figure 3.18 Output SNR after analog-to-digital converters with different number of bits.

increasing number of bits. It can be observed in Figure 3.18 that a 4-bit ADC is needed to support SC 16-QAM.

3.4.2 I/Q Mismatch

I/Q mismatch is defined as the imbalances and offsets between the I and Q branches in quadrature modulators/demodulators as shown in Figure 3.1. I/Q mismatch originates mainly in the limited layout tolerance of active and passive elements in RF transceivers. This problem is particularly serious in a direct-conversion transceiver because its quadrature modulation/demodulation frequency is much higher than that of a super-heterodyne transceiver. In designing mobile 60 GHz devices where a direct-conversion transceiver is most likely used, it is important to include I/Q mismatch in any system simulation for precise performance evaluation.

The I/Q mismatch can be divided into two parameters: gain mismatch (ε) and phase mismatch ($\Delta\varphi$). The complex baseband output signals, $y(t)$, suffering from such I/Q mismatch can be expressed as

$$y(t) = (\cos(\Delta\phi) + j\varepsilon \cdot \sin(\Delta\phi)) \cdot x(t) + (\varepsilon \cdot \cos(\Delta\phi) - j\sin(\Delta\phi)) \cdot x^*(t),$$

where $x(t)$ is input signal and x^* is the complex conjugate of x.

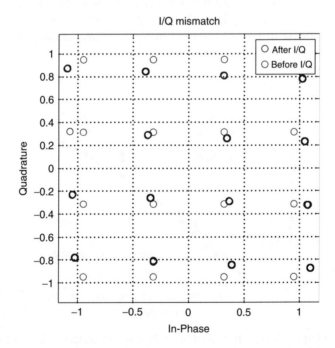

Figure 3.19 Constellations for 16-QAM signals before and after I/Q mismatch.

Figure 3.19 shows the constellations of 16-QAM signals before and after suffering from I/Q mismatch for a gain imbalance of 1 dB and a phase imbalance of 5° in I/Q branch.

References

[1] Reynolds, S.K., Floyd, B., Pfeiffer, U., Beukema, T., Grzyb, J., Haymes, C., Gaucher, B. and Soyuer, M. (2006) A silicon 60 GHz receiver and transmitter chipset for broadband communications. *IEEE Journal of Solid-State Circuits*, **41**(12), 2820–2831.

[2] Gilbert, J.M., Doan, C.H., Emami, S. and Shung, C.B. (2008) A 4-Gbps uncompressed wireless HD A/V transceiver chipset, *IEEE Micro*, **28**(2), 56–64.

[3] Choi, C.-S., Grass, E., Herzel, F., Piz, M., Schmalz, K., Sun, Y., et al. (2008) 60 GHz OFDM hardware demonstrators in SiGe BiCMOS: state-of-the-art and future development. *IEEE International Symposium on Personal, Indoor and Mobile Radio Communications*, Cannes, September.

[4] Pinel, S., Sarkar, S., Sen, P., Perumana, B., Yeh, D., Dawn D. and Laskar, J. (2008) A 90nm CMOS 60 GHz radio. *IEEE International Solid-State Circuits Conference*, San Francisco, February.

[5] Tomkins, A., Aroca, R.A., Yamamoto, T., Nicolson, S.T., Doi, Y. and Voinigescu, S.P. (2008) A zero-IF 60 GHz transceiver in 65-nm CMOS with > 3.5-Gbps links. *IEEE Custom Integrated Circuits Conference*, San Jose, CA, September.

[6] IEEE 802.15.3c (2009) *Part 15.3: Wireless Medium Access Control (MAC) and Physical Layer (PHY) Specification for High Rate Wireless Personal Area Networks (WPANs): Amendment 2: Millimeter-wave based Alternative Physical Layer Extension*.

[7] Glisic, S., Sun, Y., Herzel, F., Winkler, W., Piz, M., Grass, E., Scheytt, C. (2008) A fully integrated 60 GHz transmitter front-end with a PLL, an image rejection filter and a PA in SiGe. *European Solid-State Circuits Conference*, Edinburgh, September.

[8] Saleh, A.A.M. (1981) Frequency-independent and frequency-dependent nonlinear models of TWT amplifiers. *IEEE Transactions on Communications*, **29**, 1715–1720.

[9] Rapp, C. (1991) Effects of HPA-nonlinearity on a 4-DPSK/OFDM-signal for a digitial sound broadcasting system. *European Conference on Satellite Communications*, Liège, October.

[10] Choi, C.-S., Shoji, Y., Harada, H., Funada, R., Kato, S., Maruhashi, K. and Toyoda, I. (2006) RF impairment models for 60 GHz band system simulation. IEEE 802.15–06–0396–01–003c, IEEE 802.15.3c, 2006.

[11] Choi, C.-S., Piz, M. and Grass, E. (2009) Performance evaluation of Gbps OFDM PHY layers for 60-GHz wireless LAN applications. *IEEE International Symposium on Personal, Indoor and Mobile Radio Communications*, Tokyo, September.

[12] Osmany, S.A., Herzel, F., Schmalz, K. and Winkler, W. (2007) Phase noise and jitter modeling for fractional-N PLLs. *Advanced Radio Science*, **4**, 313–320.

[13] Herzel, F. and Piz, M. (2005) System-level simulation of noisy phase-locked loop. *European Gallium Arsenide and Other Compound Semiconductors Symposium*, Paris, October.

4

Antenna Array Beamforming in 60 GHz

Pengfei Xia

4.1 Introduction

In early 2000, the Federal Communications Commission (FCC) allocated the 57–64 GHz millimeter wave (mm-wave) band (also known as the 60 GHz frequency band) for unlicensed use, the largest contiguous radio spectrum ever allocated.[1] An equivalent isotropic radiated power (EIRP) as large as 10 watts is allowed within the band. The availability of this large frequency band with relatively loose power limitations, combined with recent advances in 60 GHz CMOS technologies, makes it attractive to support gigabit per second (Gbps) wireless applications, such as uncompressed high-definition video streaming, large file transfers, wireless gigabit Ethernet, and wireless monitors [4].

Standardization in 60 GHz wireless is under active development. The completed WirelessHD 1.0 specification is an early effort in standardizing the wireless HDMI interface, especially for consumer electronics. The IEEE 802.15.3c task group completed its specifications on 60GHz for a personal area network. In the meantime, the newly formed IEEE 802.11ad task group has also started its development of 60 GHz wireless. When completed, the IEEE 802.11ad specifications should be

---

[1]The author would like to acknowledge Samsung Information Systems America, San Jose, CA, where the work was done. Portions of the chapter are reprinted from [1–3]. Reproduced by permission © 2008 IEEE.

---

60 GHz Technology for Gbps WLAN and WPAN: From Theory to Practice
Edited by Su-Khiong (SK) Yong, Pengfei Xia and Alberto Valdes Garcia
© 2011 John Wiley & Sons, Ltd

able to support multi-Gbps throughput while at the same time being well integrated within the IEEE 802.11 family. 60 GHz wireless communications have also been actively studied for vehicular applications [5, 6].

4.2 60 GHz Channel Characteristics

To find good solutions for Gbps communications over 60 GHz, it is necessary to have a clear understanding of the underlying channel. In this section we summarize several key characteristics of the 60 GHz channel. It is also understood that the major use cases are indoor applications for consumer electronics, such as uncompressed high-definition video streaming, and large file transfer to/from a kiosk.

Channel measurements in 60 GHz show severe signal attenuation for common building materials. For example, a signal attenuation of over 20 dB is observed for a painted board, which means that signal penetration through walls would be difficult. It is thus expected that 60 GHz wireless communication for consumer electronics devices would be limited to in-room applications. Multi-room radio coverage is a problem for 60 GHz wireless that is yet to be solved. However, this brings enhanced security and decreased interference from undesirable sources.

4.2.1 Path Loss and Oxygen Absorption

In the following, we focus on 60 GHz channels within a room. One of the major challenges for mm-wave Gbps communications is the poor link budget, as radio signal propagation in the mm-wave frequency band experiences significant path loss and other degradation.

To see this, we first look at the Friis free space propagation formula,

$$P_R = P_T G_T G_R \frac{\lambda^2}{16\pi^2 R^2}, \tag{4.1}$$

where λ is the carrier wavelength, R is the distance between transmitter and receiver, G_T, G_R are the transmitter and receiver antenna gains respectively, and P_T, P_R are the transmitted and received power respectively. Clearly, the larger the carrier frequency, the smaller the carrier wavelength, and the lesser the received power. Compared with signal propagation in 2.4 GHz, the path loss in 60 GHz is thus at least 20–30 dB worse.

The 60 GHz mm-wave frequency band happens to be in an oxygen absorption band, which means that transmitted electromagnetic energy is quickly absorbed by oxygen molecules in the atmosphere. For 60 GHz wireless with a wavelength of 5 mm, a clear peak in attenuation can be observed due to the presence of oxygen. On top of the already severe path loss, oxygen absorption would further reduce the received signal power in 60 GHz wireless.

4.2.2 Multipath Fading

Most current commercial wireless communications operate with a carrier frequency smaller than 60 GHz. For example, Wi-Fi uses 2.4/5 GHz unlicensed band, the cellular system uses the 850/950 MHz and 1.8/1.9/2.1 GHz bands, and WiMAX uses the 2.3/2.5/3.5 GHz band, with carrier wavelength ranging from 8.5 to 35 cm. For those frequency bands, it is well known that the wireless channel undergoes multipath fading largely because of the scattering effect, which is the electromagnetic wave dispersion by smaller objects relative to the wavelength. Normally scattering arises due to roughness of the surface, or foliage for example.

Things change in 60 GHz, with its a much smaller wavelength of 5 mm. Radio reflection, which happens from objects that are smooth, planar and larger compared to the wavelength, instead becomes the dominant effect. This has been verified in [7], where preliminary experiments in conference room and cubicle environments show that the main propagation mechanisms for 60 GHz indoor wireless are line-of-sight (LOS), first- and second-order reflections. For indoor applications, the multipath delay spread is still significant due to relatively short reflection paths.

On the other hand, with the carrier wavelength as short as 5 mm, it becomes possible to integrate a large number of antenna elements on the whole package for both transmitter and receiver. A linear or planar antenna array may be used. An example of 36-element planar antenna array is shown in Figure 4.1, where the antenna elements are arranged uniformly on a rectangular grid.

With the availability of multiple antennas at both transmitter and receiver, multi-input multi-output (MIMO) transceivers (e.g. spatial multiplexing and/or space-time codes and/or beamforming) may be used to boost the overall channel capacity [8]. It has been reported in [9] that for 60 GHz indoor communications, a medium angular spread from 0.3 to 0.8 (with 0 being the smallest possible spread and 1 being the largest possible spread) can be observed, suggesting that space-time MIMO processing could be useful.

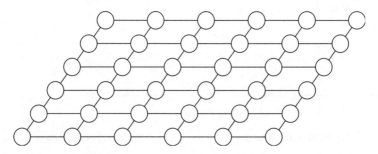

Figure 4.1 Geometry of a 36-element planar array. With wavelength as short as 5 mm, a large number of antenna elements can be placed in a reasonably sized area.

It is thus expected that 60 GHz wireless may not be able to take advantage of very high-order MIMO. Instead, lower-order MIMO techniques with one or two layers would be more appropriate for 60 GHz wireless.

Typical 60 GHz usage scenarios, such as wireless video streaming, are indoor and have little or null user mobility. In this case, the wireless channel can be assumed to be quasi-static. For this reason, closed-loop MIMO techniques such as beamforming would be preferable over the open-loop alternatives such as spatial multiplexing and/or transmit diversity.

Digital closed-loop beamforming has been used in wireless local area networks such as IEEE 802.11n, where the beamforming coefficients are applied in the digital domain (before the digital-to-analog converter (DAC) at the transmitter and after the analog-to-digital converter (ADC) at the receiver). Multiple radio frequency(RF) chains are needed though, increasing the overall cost. This is especially severe for 60 GHz wireless, where the RF cost is notoriously high. For this reason, most current 60 GHz practice uses one-dimensional (single-stream) transmit/receive beamforming. When one-dimensional beamforming is used, the receive combining operation may be implemented either in the digital domain after the ADC or in the analog domain before the ADC (see Figure 4.2). Since only one ADC is required for analog beamforming compared with multiple ADCs required for digital beamforming, analog receive beamforming is usually the choice for 60 GHz wireless.

An important benefit of using antenna array beamforming on the transmitter side is that requirements on 60 GHz power amplifiers (PAs) can be relaxed. It is reported

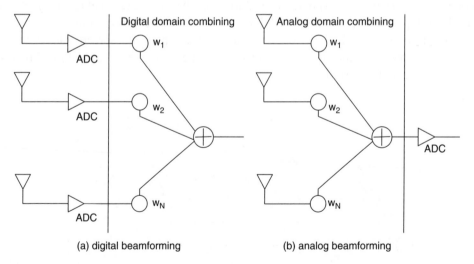

(a) digital beamforming   (b) analog beamforming

Figure 4.2 An illustration of (a) analog receive beamforming and (b) digital receive beamforming. More ADCs are required to do digital beamforming.

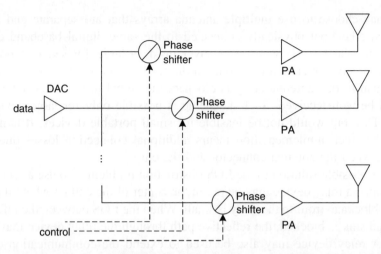

Figure 4.3 Illustration of distributed PAs in array beamforming. With multiple PAs shouldering the task, the limited gain problem for CMOS PAs can be avoided.

[10] that a gain less than 12 dB is achievable for 60 GHz CMOS PAs, which is very limited. With transmitter-side antenna array beamforming, it is possible to configure the array such that each phase shifter branch is supplied by a separate PA. The total transmit power can thus be boosted by usage of multiple (relatively) low-gain PAs. This simplifies PA design, which is very challenging at 60 GHz, and thus allows the more efficient use of PAs in the linear region without the need to exercise power backoff. Additional information on output power backoff and issues related to PAs can be found in Chapter 3 (see also Figure 4.3), and Section 6.3.2 discusses 60GHz PA implementation in silicon.

An additional benefit of antenna array beamforming is that the multipath delay spread can be significantly reduced thanks to directional transmission. Transceiver baseband design can thus be simplified considerably, allowing simpler equalization and lower resolution ADCs to be employed.

4.3 Antenna Array Beamforming

As verified in [7] through experiments in conference room and cubicle environments, the main propagation mechanisms for 60 GHz indoor wireless are LOS, first- and second-order reflections. Hence, for 60 GHz wireless, the non-line-of-sight (NLOS) paths mainly refer to the reflection paths. Since power loss generally occurs when the reflection happens and LOS is the shortest path, the LOS path is generally the path of choice. However, this is no longer true when the LOS path is obstructed, for example by human bodies or moving objects.

One may choose to use multiple antenna arrays that are separate and far apart from each other, but physically connected to the same digital baseband chip. For example, for the 60 GHz wireless receiver inside a large TV set, it is possible to place two receive antenna arrays, one on the far left and the other on the far right. If the signal to one antenna array is blocked, the signal to the other antenna array may still be well received. Such a scheme is possible only for large devices such as LCD TVs, but would not be feasible for small portable devices. It is important to note that such implementation incurs additional connection losses due to long wiring between the antenna connector and the chip.

Another possible solution to the LOS obstruction problem is to use a relay device. For example, a relay device may be put in the center of the ceiling where the chance of signal blockage from/to the relay is small. When the LOS between the information source and sink is blocked, the reflective path through the relay device may be used instead. A relay device may also be used to extend the communication coverage from within a room to across rooms. For example, a relay device may be placed in the corridor such that it is "visible" in 60 GHz to two adjacent rooms sharing a wall. 60 GHz communications between the two rooms are then possible via the relay device, which would otherwise be impossible due to severe path loss through the wall. The problem with a relay device is that it may need to be installed manually. In addition, the relay increases system latency, complexity, and cost, which is not desirable for uncompressed streaming applications and others.

To solve the NLOS problem in 60 GHz, we need to provide a mechanism that automatically finds the best path between the information source and sink, be it the LOS path or the reflective path. We also need to be able to switch to the next best path if the current best path is lost. A large antenna gain is needed for the selected path to meet the tight link budget. An antenna array is generally favored due to its ability to automatically find the best path in the changing environments. Normally two types of antenna arrays can be used: one is the fully adaptive antenna array, and the other is the switched antenna array.

A fully adaptive antenna array is the most versatile antenna array and, at least in theory, is able to form an unlimited number of beam patterns. This is achievable by configuring the phase shifters on all antenna branches with appropriate beamforming weights (see Figure 4.3). If the channel has changed (LOS being blocked, for example), a new set of beamforming weights may be configured to select the current best path. An iterative antenna training/tracking algorithm is described in Section 4.3.1 for fully adaptive antenna arrays.

A switched antenna array on the other hand assumes a structured codebook. For example a finite number of fixed, predefined beam patterns or combining strategies. By configuring the beam selection, the switched antenna array (equipped with a proper antenna training/tracking algorithm, of course) is able to automatically find the best path between the information source and sink. If the channel has changed

(LOS being blocked, for example), a new beam pattern may be selected. By taking advantage of a two-level structured codebook for switched antenna arrays, a tree-search antenna training algorithm is described in Section 4.3.2.

4.3.1　Training for Adaptive Antenna Arrays

4.3.1.1　System Description

Figure 4.4 illustrates a two-dimensional beamforming and combining system, where the beamforming and combining operation take place in the analog domain [1]. The one-dimensional transmit and receive beamforming studied in [2], where only one RF chain is used with a single transmit beamforming vector and a single receive combining vector, is simply a special case of Figure 4.4. As it turns out, the antenna training algorithm therein would also be a degenerate version of the more general algorithm to be described here. Another motivation for considering a more general two-dimensional system here is that, as shown in [7, 9], rank-2 transmission is indeed a viable solution for 60 GHz indoor applications.

In particular, transmit beamforming is achieved via the two independent beamforming vectors $\{v_{1,i}\}_{i=1}^{N_t}$, $\{v_{2,i}\}_{i=1}^{N_t}$, where $v_{1,i}$ and $v_{2,i}$ represent the ith beamforming coefficients on the ith antenna for the first and second stream respectively. On the other hand, receive combining is achieved via the two independent beamforming vectors $\{u_{1,j}\}_{j=1}^{N_r}$, $\{u_{2,j}\}_{j=1}^{N_r}$ where $u_{1,j}$ and $u_{2,j}$ represent the jth receive combining coefficients on the jth antenna for the first and second stream respectively. Here N_t and N_r are the number of transmit and receive antennas respectively. In this chapter, bold-face upper- and lower-case letters denote matrices and column vectors, respectively. $(\cdot)'$ denotes the Hermitian transpose of a matrix/vector.

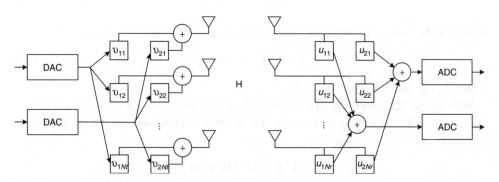

Figure 4.4　Illustration of a two-dimensional beamforming system in 60 GHz. Two transmit beamforming vectors and two receive combining vectors are used to support rank-2 transmissions simultaneously [1]. Reproduced by permission of © 2008 IEEE.

The input–output relationship in Figure 4.4 can be expressed in matrix form as

$$\mathbf{y} = \mathbf{Hx} + \mathbf{n}, \tag{4.2}$$

where $\mathbf{H}$ is the $N_r \times N_t$ channel matrix between the transmitter and the receiver, and $\mathbf{y}, \mathbf{x}, \mathbf{n}$ are the $N_r \times 1$, $N_t \times 1$, $N_r \times 1$ received signal vector, transmitted signal vector and the additive Gaussian noise vector, respectively [1].

For simplicity, we consider a flat fading channel model here.

4.3.1.2 Optimal Solution: Eigenbeamforming

We are interested in acquiring the optimal beamforming coefficients on both the transmitter and receiver side. Let us now assume that knowledge of $\mathbf{H}$ is available at both transmitter and receiver thanks to slow channel variation. The optimal transmitter and receiver can then be obtained through singular value decomposition (SVD) of the channel [11]. In particular, let $\mathbf{H} = \mathbf{U\Sigma V'}$ where $\mathbf{U}$ is the $N_r \times N_r$ left singular matrix, $\mathbf{V}$ is the $N_t \times N_t$ right singular matrix of $\mathbf{H}$, and $\mathbf{\Sigma}$ is the $N_r \times N_t$ diagonal matrix containing all the singular values associated with $\mathbf{H}$ in a non-increasing order, $\sigma_1 > \sigma_2 > \cdots > \sigma_{\min(N_t, N_r)}$. Notice that both $\mathbf{U}, \mathbf{V}$ are unitary matrices. In fact,

$$\begin{aligned} \mathbf{HH'} &= \mathbf{U\Sigma^2 U'}, \\ \mathbf{H'H} &= \mathbf{V\Sigma^2 V'}, \end{aligned} \tag{4.3}$$

that is, $\mathbf{U}$ is simply the eigenmatrix of $\mathbf{HH'}$, and $\mathbf{V}$ is simply the eigenmatrix of $\mathbf{H'H}$. Let N be the number of RF chains, where $N = 1$ leads to one-dimensional beamforming and $N = 2$ is illustrated in Figure 4.4. It suffices to precode at the transmitter using

$$\tilde{\mathbf{V}} = [\mathbf{v}_1, \cdots, \mathbf{v}_N], \tag{4.4}$$

which is the first N columns of $\mathbf{V}$, and combine at the receiver using

$$\tilde{\mathbf{U}} = [\mathbf{u}_1, \cdots, \mathbf{u}_N], \tag{4.5}$$

which is the first N columns of $\mathbf{U}$.

With linear beamforming operation of $\tilde{\mathbf{V}}$ at the transmitter and linear combining operation of $\tilde{\mathbf{U}}'$ at the receiver side, the MIMO channel $\mathbf{H}$ can be diagonalized. In particular,

$$\begin{aligned} \mathbf{z} = \tilde{\mathbf{U}}'\mathbf{y} &= \tilde{\mathbf{U}}'(\mathbf{Hx} + \mathbf{n}) \\ &= \tilde{\mathbf{U}}' \left(\mathbf{U\Sigma V'} \cdot \tilde{\mathbf{V}}\mathbf{s} + \mathbf{n} \right) \\ &= \tilde{\mathbf{\Sigma}}\mathbf{s} + \tilde{\mathbf{U}}'\mathbf{n}, \end{aligned} \tag{4.6}$$

where $\mathbf{x} = \tilde{\mathbf{V}}\mathbf{s}$ indicates that the original information signal vector $\mathbf{s}$ of size $N \times 1$ is precoded by $\tilde{\mathbf{V}}$, and $\tilde{\mathbf{\Sigma}}$ is the first $N \times N$ submatrix of $\mathbf{\Sigma}$ on the upper-left corner and is itself a diagonal matrix. Along with the channel diagonalization comes equalizer simplicity, that is, the multiple streams are not interfering with each other at the receiver side, thus leading to simple per-stream equalization. Notice that the additive channel noise is not enhanced thanks to the unitary nature of $\tilde{\mathbf{U}}$ [1].

Conventionally, the channel matrix $\mathbf{H}$ needs to be estimated to compute the SVD at the receiver side and furthermore the beamformer (a beamformer is simply a beamforming vector) coefficients $\tilde{\mathbf{V}}$ need to be fed back. This task of direct estimation and feedback is manageable when the number of antennas is small – for example, $N_t, N_r \leq 4$ as in IEEE 802.11n systems. However, the computation complexity and training overhead quickly get out of control when the number of antennas N_t, N_r grows large. For wireless communications over the 60 GHz frequency band, we are mostly interested in $N \ll N_t, N \ll N_r$. In such cases, the conventional direct-estimate-and-feedback approach has a computational complexity and training overhead of the order of $\min^2(N_t, N_r) \times \max(N_t, N_r)$, and is neither bandwidth-efficient nor computation-efficient.

4.3.1.3 Power Iteration Principle

Having discussed the inefficiency of the conventional direct-estimate-and-feedback approach, we now introduce the so-called principle of power iteration, which holds for general matrix $\mathbf{H}$ [12] and is useful for our mm-wave beamforming protocol development. Consider the SVD decomposition of $\mathbf{H}$, which can be rewritten as

$$\mathbf{H} = \sigma_1 \mathbf{u}_1 \mathbf{v}_1' + \sigma_2 \mathbf{u}_2 \mathbf{v}_2' + \cdots + \sigma_P \mathbf{u}_P \mathbf{v}_P', \tag{4.7}$$

where $P = \min(N_t, N_r)$. Furthermore, we have

$$\begin{aligned}
\mathbf{H}\mathbf{H}' &= \sigma_1^2 \mathbf{u}_1 \mathbf{u}_1' + \sigma_2^2 \mathbf{u}_2 \mathbf{u}_2' + \cdots + \sigma_P^2 \mathbf{u}_P \mathbf{u}_P', \\
\mathbf{H}'\mathbf{H} &= \sigma_1^2 \mathbf{v}_1 \mathbf{v}_1' + \sigma_2^2 \mathbf{v}_2 \mathbf{v}_2' + \cdots + \sigma_P^2 \mathbf{v}_P \mathbf{v}_P',
\end{aligned} \tag{4.8}$$

using the fact that

$$\begin{aligned}
\mathbf{u}_i' \mathbf{u}_j &= \delta_{ij}, \\
\mathbf{v}_i' \mathbf{v}_j &= \delta_{ij},
\end{aligned} \tag{4.9}$$

in which δ_{ij} is the Kronecker delta. For notational simplicity, we define an even power of $\mathbf{H}$ as

$$\mathbf{H}^{2m} := \underbrace{\mathbf{H}'\mathbf{H} \times \mathbf{H}'\mathbf{H} \cdots \times \mathbf{H}'\mathbf{H}}_{m \ pairs}, \tag{4.10}$$

where

$$\mathbf{H}^{2m} = \sigma_1^{2m}\mathbf{v}_1\mathbf{v}_1' + \sigma_2^{2m}\mathbf{v}_2\mathbf{v}_2' \cdots + \sigma_P^{2m}\mathbf{v}_P\mathbf{v}_P' \qquad (4.11)$$

thanks to (4.8) and (4.9). Here m is any positive integer.

An important observation is that, as m increases, $\sigma_i^{2m}/\sigma_1^{2m}$ decreases in an exponential manner, for all $i = 2, \ldots, P$. This means that the contributions to the matrix product $\mathbf{H}^{2m}$ from the last $P - 1$ non-principal terms are shrinking, while the contribution from the principal term, $\sigma_1^{2m}\mathbf{v}_1\mathbf{v}_1'$, is dominant. Mathematically,

$$\lim_{m\to\infty} \mathbf{H}^{2m} = \sigma_1^{2m}\mathbf{v}_1\mathbf{v}_1'. \qquad (4.12)$$

We define an odd power of $\mathbf{H}$ similarly as

$$\mathbf{H}^{2m+1} := \mathbf{H} \times \underbrace{\mathbf{H}'\mathbf{H} \times \mathbf{H}'\mathbf{H} \cdots \times \mathbf{H}'\mathbf{H}}_{m\ pairs}, \qquad (4.13)$$

where

$$\mathbf{H}^{2m+1} = \sigma_1^{2m+1}\mathbf{u}_1\mathbf{v}_1' + \sigma_2^{2m+1}\mathbf{u}_2\mathbf{v}_2' + \cdots + \sigma_P^{2m+1}\mathbf{u}_P\mathbf{v}_P' \qquad (4.14)$$

thanks to (4.8), and (4.9). Similarly,

$$\lim_{m\to\infty} \mathbf{H}^{2m+1} = \sigma_1^{2m+1}\mathbf{u}_1\mathbf{v}_1'. \qquad (4.15)$$

The good news is that the convergence in Equations (4.12) and (4.15) is usually achieved with a small value of $m = 3$ or 4 [1]. This is clearly illustrated in Figure 4.5 where a snapshot of the relative strength $\sigma_i^{2m}/\sigma_1^{2m}$ is shown, indicating that the contribution from $\sigma_i^{2m}/\sigma_1^{2m}$ diminishes sharply as the power index m increases. For example, for $m = 4$, $\sigma_2^{2m}/\sigma_1^{2m} < 1\%$ already.

We next present a multi-stage iterative antenna training protocol for acquiring the antenna beamforming coefficients at both transmitter and receiver, where the number of stages depends on the number of transmit beamforming (receive combining) vectors that need to be trained. Without loss of generality, we present a case with two training stages; the analysis can be extended to an arbitrary number of stages.

In this chapter, we focus on time division duplexing transmissions where the uplink and downlink channels are mutually reciprocal, that is, the downlink channel is the Hermitian transpose of the uplink channel:

$$\mathbf{H}_{downlink} = \mathbf{H}_{uplink}'. \qquad (4.16)$$

4.3.1.4 Iterative Antenna Training–Stage 1

The purpose of the first stage is to acquire the first beamforming vectors for transmitter and receiver, i.e. the first column in Equations (4.4) and (4.5). In the following,

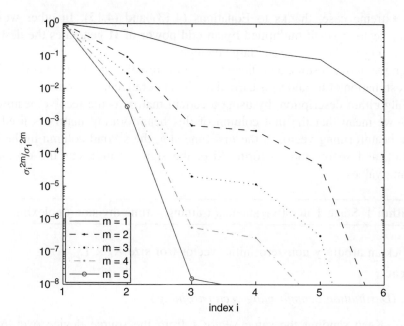

Figure 4.5 $\sigma_i^{2m}/\sigma_1^{2m}$ diminishes sharply as m increases, demonstrating that the power iteration converges quickly, typically after 3–4 iterations [1]. Reproduced by permission of © 2008 IEEE.

t represents the interim transmit beamforming vector and $\mathbf{r}$ represents the interim receive beamforming vector.

Notice that, for arbitrary $N_t \times 1$ vector $\mathbf{t} = \sum_{i=1}^{P} c_i \mathbf{v}_i$, where $c_i \mathbf{v}_i$ is the projection of $\mathbf{t}$ on the direction of $\mathbf{v}_i$, it can be readily verified that

$$
\begin{aligned}
\mathbf{H}^{2m}\mathbf{t} &= \sigma_1^{2m}\mathbf{v}_1\mathbf{v}_1' \cdot \left(\sum_{i=1}^{P} c_i \mathbf{v}_i\right) \\
&= c_1 \sigma_1^{2m} \mathbf{v}_1
\end{aligned}
\tag{4.17}
$$

in the extreme case, thanks to Equations (4.11) and (4.12). It is assumed here, for notational simplicity, that the convergence in Equation (4.12) has already been achieved. An arbitrary vector $\mathbf{t}$ when multiplied by an even power of $\mathbf{H}$ generates the source beamforming vector $\mathbf{v}_1$, if m is large enough and the projection of $\mathbf{t}$ on $\mathbf{v}_1$ is not zero, which is true with probability 1. Similarly,

$$
\begin{aligned}
\mathbf{H}^{2m+1}\mathbf{t} &= \sigma_1^{2m+1}\mathbf{u}_1\mathbf{v}_1' \cdot \left(\sum_{i=1}^{P} c_i \mathbf{v}_i\right) \\
&= c_1 \sigma_1^{2m+1} \mathbf{u}_1
\end{aligned}
\tag{4.18}
$$

in the extreme case, thanks to Equations (4.13) and (4.15). In other words, an arbitrary vector **t** when multiplied by an odd power of **H** generates the destination beamforming vector $\mathbf{u}_1$, if m is large enough.

Given the above results, an efficient way of estimating $\mathbf{u}_1$, $\mathbf{v}_1$ is via Algorithm 1, where estimation of $\mathbf{u}_1$ and $\mathbf{v}_1$ is carried out in an iterative manner until convergence. In the algorithm description, by using a certain matrix as the receive beamforming matrix, we mean that the first column of this beamforming matrix is used as the receive beamforming vector in the first time slot, the second column in the second time slot, and so on and so forth. Also, the sign $\leftarrow$ indicates the operation of assigning values.

Algorithm 1 Stage 1 iterative antenna training – time division duplexing

---

0. Pick an arbitrary non-zero initial vector **t** of size $N_t \times 1$.

repeat

   1. (*Destination beamforming vector training*)

   1.1. Keep sending the same vector **t** from the source device over N_r time slots, while at the same time using $\mathbf{I}_{N_r}$ as the receive beamforming matrix at the destination device.

   1.2. Update the vector **r** by the received vector at the destination device, i.e.

$$\mathbf{r} \leftarrow \mathbf{Ht} + \text{noise}_1 \tag{4.19}$$

   1.3. Normalize the vector **r** to be of unit norm.

   2. (*Source beamforming vector training*)

   2.1. Keep sending the same vector outcome **r** from the destination device over N_t time slots while at the same time use $\mathbf{I}_{N_t}$ as the receive beamformer at the source device.

   2.2. Update the vector **t** by the received vector at the source device, i.e.

$$\mathbf{t} \leftarrow \mathbf{H}'\mathbf{r} + \text{noise}_2 \tag{4.20}$$

   2.3. Normalize the vector **t** to be of unit norm.

until a predefined number of iterations is reached

The iterative method works by repeating steps (1) and (2) m times. Notice that vector normalization is needed every time an updated vector of **t** or **r** is transmitted. Thanks to Equations (4.17) and (4.18), vector **r** at the destination device converges

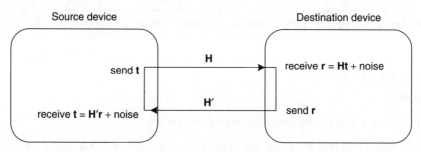

Figure 4.6 Illustration of the stage 1 iterative antenna training algorithm 1. Typically the algorithm converges after 3–4 iterations [1]. Reproduced by permission of © 2008 IEEE.

to the desired singular vector $\mathbf{u}_1$, while vector $\mathbf{t}$ at the source device converges to the desired singular vector $\mathbf{v}_1$. An illustration of the algorithm is shown in Figure 4.6.

As we can see, the purpose of using $\mathbf{I}_{N_r}$ as the receive beamforming matrix in step 1.1 is to form Equation (4.19) in a straightforward manner. Similarly, the purpose of using $\mathbf{I}_{N_t}$ as the transmit beamforming matrix in step 2.1 is to form Equation (4.20) in a straightforward manner.

4.3.1.5 Iterative Antenna Training–Stage 2

In stage 2, we are interested in acquiring the second beamforming vector for transmitter and receiver, i.e. the second column in Equations (4.4) and (4.5). If we compare the antenna training process to onion peeling, the iterative method in stage 1 (2) is like peeling the first (second) layer of an onion.

Let $\mathbf{t}_\varnothing$ be the projection of $\mathbf{t}$ (the same arbitrary vector picked in the first stage) onto the null space of $\mathbf{v}_1$, where

$$\mathbf{t}_\varnothing = \mathbf{t} - (\mathbf{v}_1'\mathbf{t})\mathbf{v}_1 = \sum_{i=2}^{P} c_i\mathbf{v}_i. \tag{4.21}$$

Notice that $\mathbf{t}_\varnothing$ contains contributions from all layers but v_1 due to the projection operation.

It is straightforward that

$$\mathbf{H}^{2m}\mathbf{t}_\varnothing = \left(\sum_{i=1}^{P} \sigma_i^{2m}\mathbf{v}_i\mathbf{v}_i'\right)\left(\sum_{i=2}^{P} c_i\mathbf{v}_i\right)$$

$$= \sum_{i=2}^{P} c_i\sigma_i^{2m}\mathbf{v}_i, \tag{4.22}$$

Algorithm 2 Stage 2 iterative antenna training – time division duplexing

---

0. Use the same non-zero initial vector $\mathbf{t}$ from stage 1 and find $\mathbf{t}_\varnothing$, its projection onto the null space of $\mathbf{v}_1$.

repeat

    1. (*Destination beamforming vector training*)

    1.1. Keep sending $\mathbf{t}_\varnothing$ from the source device over N_r time slots while at the same time using $\mathbf{I}_{N_r}$ as the receive beamforming matrix at the destination device.
    1.2. Update the vector $\mathbf{r}$ by the received vector at the destination device, i.e.

$$\mathbf{r} \leftarrow \mathbf{H}\mathbf{t}_\varnothing + \text{noise}_1. \tag{4.23}$$

    1.3. Project the received vector $\mathbf{r}$ on the null space of $\mathbf{u}_1$ yielding its projection

$$\mathbf{r}_\varnothing \leftarrow \mathbf{r} - (\mathbf{u}_1'\mathbf{r})\mathbf{u}_1. \tag{4.24}$$

    1.4. Normalize the projected vector $\mathbf{r}_\varnothing$ to be of unit norm.

    2. (*Source beamforming vector training*)

    2.1. Keep sending the vector outcome $\mathbf{r}_\varnothing$ in (4.24) from the destination device over N_t time slots while at the same time using $\mathbf{I}_{N_t}$ as the receive beamformer at the source device.
    2.2. Update the vector $\mathbf{t}$ by the received vector, i.e.

$$\mathbf{t} \leftarrow \mathbf{H}'\mathbf{r}_\varnothing + \text{noise}_2 = \mathbf{H}^2\mathbf{t}_\varnothing + \text{noise}_c, \tag{4.25}$$

where noise_c is the composite noise due to noise_1 and noise_2.
    2.3. Project $\mathbf{t}$ on the null space of $\mathbf{v}_1$ yielding its projection

$$\mathbf{t}_\varnothing \leftarrow \mathbf{t} - (\mathbf{v}_1'\mathbf{t})\mathbf{v}_1. \tag{4.26}$$

    2.4. Normalize the vector $\mathbf{t}_\varnothing$ to be of unit norm.
until a predefined number of iterations is reached

---

and that

$$\mathbf{H}^{2m+1}\mathbf{t}_\varnothing = \left(\sum_{i=1}^{P} \sigma_i^{2m+1}\mathbf{u}_i\mathbf{v}_i'\right)\left(\sum_{i=2}^{P} c_i\mathbf{v}_i\right)$$

$$= \sum_{i=2}^{P} c_i\sigma_i^{2m+1}\mathbf{u}_i. \tag{4.27}$$

Using the same argument as in Equations (4.12) and (4.15), we arrive at

$$\lim_{m \to \infty} \mathbf{H}^{2m} \mathbf{t}_\varnothing = c_2 \sigma_2^{2m} \mathbf{v}_2,$$

$$\lim_{m \to \infty} \mathbf{H}^{2m+1} \mathbf{t}_\varnothing = c_2 \sigma_2^{2m+1} \mathbf{u}_2. \tag{4.28}$$

$\mathbf{t}_\varnothing$, the projection of arbitrary non-zero vector $\mathbf{t}$ onto the null space of $\mathbf{v}_1$, generates the nth source beamforming vector $\mathbf{v}_2$ when multiplied by $\mathbf{H}^{2m}$, and generates the nth destination beamforming vector $\mathbf{u}_2$ when multiplied by $\mathbf{H}^{2m+1}$, if m is large enough.

Given the above results, an efficient way of estimating $\mathbf{u}_2$, $\mathbf{v}_2$ is via Algorithm 2, where estimation of $\mathbf{u}_2$ and $\mathbf{v}_2$ is carried out in an iterative manner until convergence is achieved. In particular, the iterative method works by repeating step (1) and (2) of the algorithm. By Equation (4.28), $\mathbf{t}_\varnothing$ converges to the desired source beamforming vector $\mathbf{v}_2$ while $\mathbf{r}_\varnothing$ converges to the desired destination beamforming vector $\mathbf{u}_2$. An illustration of the algorithm is given in Figure 4.7.

Unique to the inner stages are the steps of Equations (4.24) and (4.26) in carrying out the null space projection. We would like to emphasize here the critical role that the projection operation plays. Suppose we work in stage 2 but the projection operations Equations (4.24) and (4.26) are skipped. Instead, the vector $\mathbf{r}$ in Equation (4.23) and vector $\mathbf{t}$ in Equation (4.25) are directly used in later iterations as in stage 1. It is easy to realize that Equation (4.23) is identical to Equations (4.19) and (4.25) is identical to Equation (4.20). We conclude that the resultant vectors would converge again to $\mathbf{u}_1$ and $\mathbf{v}_1$ which are already obtained in stage 1, but not the desired output $\mathbf{u}_2$, $\mathbf{v}_2$ in stage 2. In other words, the interim source beamforming vector $\mathbf{H}^{2m} \mathbf{t}_\varnothing$ and the interim destination beamforming vector $\mathbf{H}^{2m+1} \mathbf{t}_\varnothing$ shall fall onto the null space of $\mathbf{v}_1$ and null space of $\mathbf{u}_1$ respectively.

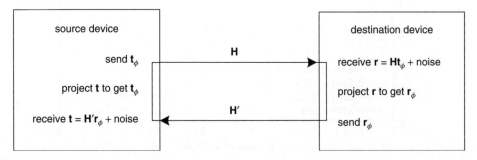

Figure 4.7 Illustration of the stage 2 iterative antenna training algorithm. Typically the algorithm converges after 3–4 iterations. Notice that the null space projection operation is needed for each interim transmit/receive beamforming vector calculation [1]. Reproduced by permission of © 2008 IEEE.

Despite many indoor channel measurements and modeling efforts for the 5 GHz WLAN band and 3–10 GHz UWB band, there are very limited channel measurements and modeling available for the 60 GHz frequency band. IEEE 802.15.3c proposed a generic 60 GHz channel model based on the angular-domain-extended Saleh–Valenzuela model [13]. We further employ the generic MIMO model with clustering information [14] to generate the correlated MIMO channel. Particularly in the simulations, we assume the existence of two clusters, with mean angle of departure of 30° and 50° respectively from the transmitter side, mean angle of arrival of 30° and 80° respectively at the receiver side, angle spread of 30° for both clusters, and even power distribution between the two clusters. With these and input parameters from [13], we are able to generate the correlated channel matrix for simulation purposes. We further limit ourselves to two-dimensional transmit precoding and receive combining in a MIMO setup of $N_t = 12$, $N_r = 11$ antennas.

Figures 4.8 and 4.9 illustrate the convergence behavior for stage $n = 1, 2$ respectively. The dashed curve represents respectively the achievable channel gain upper-bound σ_n over the first and second eigenmodes, which are available through perfect SVD in a noiseless environment. For each eigenmode, the achieved channel

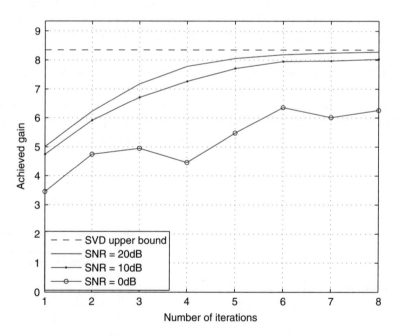

Figure 4.8 Convergence behavior for the first eigenmode. The operation converges only if the operating SNR is large enough [1]. Reproduced by permission of © 2008 IEEE.

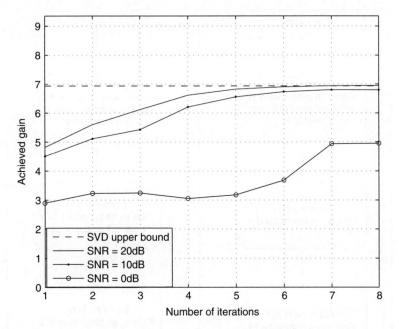

Figure 4.9 Convergence behavior for the second eigenmode. The operation converges only if the operating SNR is large enough [1]. Reproduced by permission of © 2008 IEEE.

gains $|\mathbf{u}'_n \mathbf{H} \mathbf{v}_n|$ with the actually estimated beamforming vectors $\mathbf{u}_n, \mathbf{v}_n$ are plotted against number of iterations for different signal-to-noise ratio (SNR) scenarios (corresponding to different noise variances). For comparison purposes, the same channel realization and noise realization are used for all different SNRs. Although only one channel/noise realization is used, the same behavior occurs for other realizations in general.

An important observation is that, for high working SNR, the iterations converge very quickly to or near the upper bound in a few iterations. Typically, three or four iterations are enough to capture most of the available gains. However, when the operating SNR is low, the iteration either converges slowly to far below the upper bound, or does not converge at all. It is thus important to have the iterative training method work in a relatively medium to high SNR scenario. Typically pseudo-random sequence spreading can be used to improve the working SNR.

A complete flowchart is given in Figure 4.10 to illustrate the multistage iterative antenna training method.

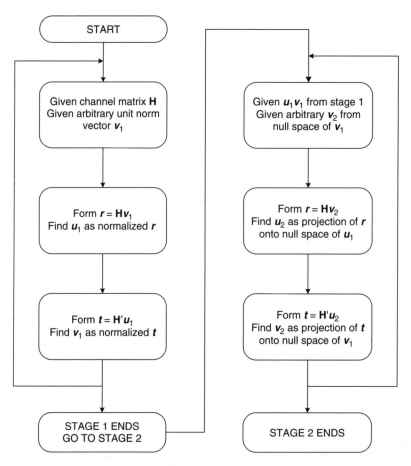

Figure 4.10 Illustration of the multi-stage iterative antenna training in stage 1 and 2. Notice that stage 1 results v_1, u_1 are to be used in stage 2 for null space projection.

4.3.1.6 Tracking for Adaptive Antenna Arrays

In practical systems featuring adaptive antenna arrays, the actual communication is usually divided into two phases, the antenna training phase where beamforming weighting coefficients are calculated for both transmitter and receiver to achieve as large a beamforming gain as possible, and the payload transmission phase where actual data is communicated using the beamformed high-gain link. It is generally desirable that the antenna training phase be as short as possible. This is because short antenna training means lower overhead and thus larger effective throughput, and long antenna training may violate the underlying assumption that the wireless channel does not change much from the antenna training phase to the payload transmission phase.

The above antenna training algorithm can be used to acquire proper beamforming coefficients to establish a high-gain beamformed link between transmitter and receiver, without any prior channel/direction information at all. On the other hand, since the wireless channel is slowly changing, it is thus necessary for the transmitter and receiver to track the channel changes by changing their beamforming coefficients accordingly. Nevertheless the complete training algorithm would be an overkill for the tracking purpose. Instead, it is enough to use only one iteration of the training algorithm for each stage for antenna tracking purposes. Tracking is typically performed once in a certain number of super-frames.

4.3.2   Training for Switched Antenna Arrays

Section 4.3.1 describes an iterative antenna training/tracking algorithm for fully adaptive antenna arrays. However, such an iterative algorithm could not applied if a switched antenna array were used. Nevertheless, by taking advantage of the codebook structure for switched antenna arrays, a simpler antenna training algorithm may be developed as follows.

It is assumed that each transmitter/receiver maintains a two-level codebook structure, which can best be explained through the example in Figure 4.11. Specifically, the transmitter (and receiver) maintains two codebooks – a sector codebook and a beamformer codebook. Remember that for switched antenna arrays, a codebook is simply a collection of fixed, predefined beamforming vectors (or combining strategies).

A sector codebook is a collection of coarse sectors, with each sector covering a broad direction in the space, and all sectors together covering the entire space of interest. Here the space may represent the space of azimuth angles, elevation angles or both. As illustrated in Figure 4.11(a), the transmitter-side sector codebook simply consists of two sectors that jointly cover the entire space of elevation angles. Similarly, a sector codebook may be maintained at the receiver side as well (Figure 4.11(b)). Notice that the size of transmitter-side sector codebook is not necessarily the same as that of receiver-side sector codebook.

A beamformer codebook is a collection of fine beams with each beam covering a fine direction in the space, and all beams together covering the entire space of elevation angles. As illustrated in Figure 4.11(c), the transmitter-side beam codebook consists of six beams that jointly cover the entire space (all possible elevation angles). Similarly, a beam codebook may be maintained at the receiver side as well (Figure 4.11(d)).

Without loss of generality, each sector in the sector codebook corresponds to several beams in the beam codebook. For example, sector 1 corresponds to beams 1–3, meaning that beams 1–3 jointly have similar coverage to sector 1 in the space

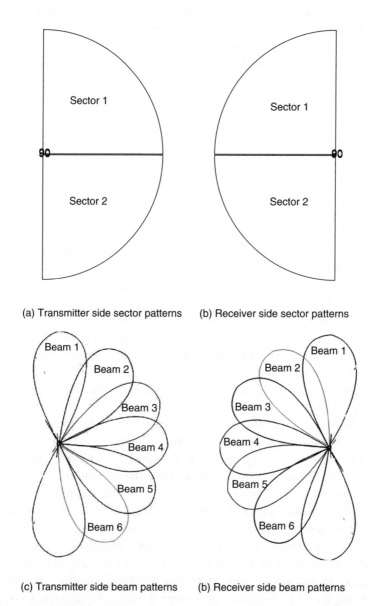

(a) Transmitter side sector patterns   (b) Receiver side sector patterns

(c) Transmitter side beam patterns   (b) Receiver side beam patterns

Figure 4.11 Illustration of the two-level codebook structure. In this example, the coarse sector codebook contains two sectors covering the whole space while the fine beam codebook contains six beams covering the same space. Furthermore, sector 1 has roughly the same coverage as beams 1–3 together, while sector 2 has roughly the same coverage as beams 4–6 together.

considered. Similarly, beams 4–6 jointly have roughly similar coverage to sector 2. Notice that here the "coverage" focuses more on which direction covered, rather than how far it is covered.[2] It is implied here that the transmitter switched antenna array is able to physically form each candidate sector pattern in the transmit sector codebook, as well as each candidate beam pattern in the transmit beam codebook. It is also implied that the receiver switched antenna array is able to physically form each candidate sector pattern in the receive sector codebook, as well as each candidate beam pattern in the receive beam codebook.

With the two-level codebooks maintained as such for the transmitter and receiver switched antenna arrays, a tree search antenna training method can be carried out as in Algorithm 3. In the initialization stage, the transmitter and receiver need to exchange control signaling so that the two-level transmitter-side codebook information (such as number of transmit sectors and number of transmit beams) is known by the receiver, and similarly the two-level receiver-side codebook information (such as number of receive sectors and number of receive beams) is known by the transmitter. Notice that this information exchange needs to be carried out reliably and in every direction. This usually can be done via a relative low-rate physical layer mode which has almost omnidirectional transmissions.

Algorithm 3 A tree search training method

---

0. *Initialization*

0.1. Maintain a coarse sector codebook and a fine beam codebook at the transmitter.
0.2. Maintain a coarse sector codebook and a fine beam codebook at the receiver.
0.3. Transmitter signals the transmitter-side codebook information to the receiver.
0.4. Receiver feeds back the receiver-side codebook information to the transmitter.

1. *Coarse sector training*

1.1. For each possible pair of transmit sector i and receive sector j, transmit a training sequence with sector i and receive with sector j, and record the SNR as $\rho(i, j)$.
1.2. Receiver selects the best pair of transmit sector i^* and receive sector j^* such that the corresponding SNR is the largest.
1.3. Receiver feeds back the transmitter-side sector index i^*.

---

[2]In general, a fine beam covers farther than a broad sector with the same given power.

2. *Fine beam training*

2.1. For each possible pair of transmit beam p within the coverage of sector i^* and receive beam q within the coverage of sector j^*, transmit a training sequence with beam p and receive with beam q, and record the SNR as $\rho(i^*, j^*, p, q)$.
2.2. Receiver selects the best pair of transmit beam p^* and receive beam q^* such that the corresponding SNR is the largest.
2.3. Receiver feeds back the transmitter-side beam index p^*.

3. Transmit beam p^* and receive beam q^* are to be used for data transmissions.

The overall idea is divide and conquer, and is clearly illustrated in Figure 4.12. In principle, one may perform a brute force search over all possible pairs of transmit and receive sectors, while the best pair is selected after the first-round brute force search. In the example in Figure 4.12, transmit sector 1 and receive sector 1 are determined as the best pair of sectors. Brute force search can be terminated if the required performance such as SNR/SINR between two RX/TX sector pairs is reached.

The search is then brought to the beam level. Within the coverage of the selected pair of transmit/receive sectors earlier, a constrained search over all possible pairs of transmit and receive beams is performed and the best pair is selected after the second-round constrained search. In the example in Figure 4.12, transmit beam 2 (within the coverage of transmit sector 1) and receive beam 3 (within the coverage of receive sector 1) are found to be the best pair of beams. Transmit beam 2 and receive beam 3 are then used to form the high-gain link, over which the beamformed transmission takes place.

Notice that the tree search algorithm can be easily extended to multiple levels, if a multilevel codebook structure is assumed by transmitter and receiver. In some cases, multiple best candidates may be preferred. For example, if the best candidate (of transmit and receive beam) is blocked, the second best candidate can be used right off the shelf without a new complete tree search from the bottom. Algorithm 3 can also be easily adapted to find a short list of best candidates for such usage.

4.3.3 Channel Access in 60 GHz Wireless Networks

High-speed data packets in 60 GHz networks may access the wireless medium in a conventional time division manner. This can be done either following a reservation based time division multiple access (TDMA) protocol, or a carrier sensing multiple access/collision avoidance (CSMA/CA) protocol, or a hybrid of the two. Notice that high-speed data transmissions in 60 GHz would typically be directional as a natural result of transmit beamforming. For this reason, it would be very beneficial

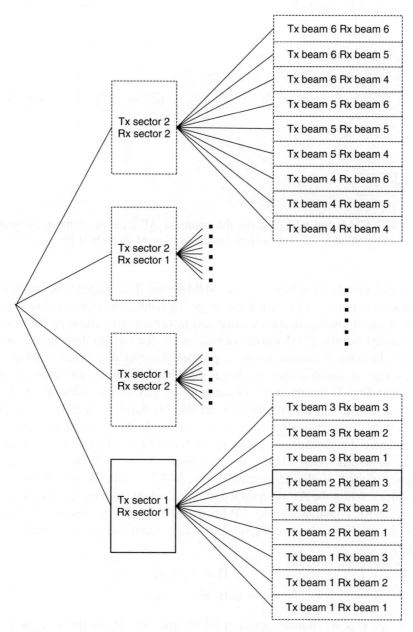

Figure 4.12   A tree search training algorithm for switched antenna arrays.

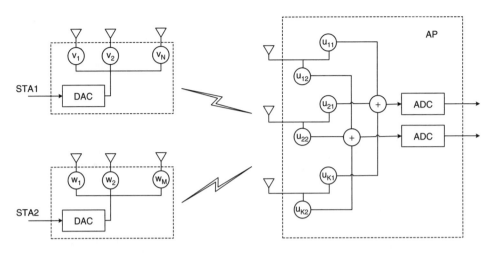

Figure 4.13 STA1 and STA2 access the common AP over the same time-frequency resources using SDMA [2]. Reproduced by permission of © 2008 IEEE.

to do spatial division multiple access (SDMA) for data packets, so as to extend communication range and network coverage, to reduce interference from adjacent users, to mitigate multipath interference and to reduce network-wide RF pollution.

It is realized that the TDMA protocol may not be the best multiple access protocol for highly directional transmissions, as pointed out by Cooper and Goldburg [15]. A general rule of thumb is that, the larger the desired beamforming gain, the higher the directionality. Typically, a total beamforming gain of the order of 20–30 dB is required to achieve Gbps throughput in the 60 GHz band. Hence, the transmissions would be highly directional and it is better to use SDMA. In this section, we are more interested in a multiple access system where STA1 and STA2 access a common AP at the same time using SDMA, as illustrated in Figure 4.13. In particular, STA1 is equipped with an N-element antenna array, STA2 is equipped with an M-element antenna array, while the AP is equipped with a K-element antenna array. Each STA has a single RF chain, while the AP has two RF chains to support simultaneous transmissions to/from the two STAs. The uplink input–output relationship can be written as

$$
\begin{aligned}
y_1 &= \mathbf{u}_1' \mathbf{H}_1 \mathbf{v} \cdot s_1 + n_1, \\
y_2 &= \mathbf{u}_2' \mathbf{H}_2 \mathbf{w} \cdot s_2 + n_2,
\end{aligned}
\tag{4.29}
$$

where $\mathbf{H}_1$ is $K \times N$ channel between STA1 and AP, $\mathbf{H}_2$ is the $K \times M$ channel between STA2 and AP, $\mathbf{u}_1 = [u_{11}, \cdots, u_{K1}]^T$ is the receive beamforming vector for STA1, $\mathbf{u}_2 = [u_{12}, \cdots, u_{K2}]^T$ is the receive beamforming vector for STA2, $\mathbf{v}, \mathbf{w}$ are transmit beamforming vectors for STA1 and STA2 respectively, s_1, s_2 are information symbols originated from STA1 and STA2 respectively, y_1, y_2 are received symbols at the output of the two ADCs respectively, and n_1, n_2 are independent additive white Gaussian random variables respectively for each STA.

In determining the beamforming vectors $\mathbf{u}_1, \mathbf{u}_2, \mathbf{v}, \mathbf{w}$, we face a similar situation as in the point-to-point systems, that is, the number of available antenna elements (N, M or K) would be significantly larger than the number of RF chains. The conventional estimate-and-feedback approach would thus be very inefficient, as the training overhead and complexity are proportional to the number of antenna elements. We are thus motivated to follow the similar power iteration principle in acquiring the beamforming coefficients for STAs and the AP. Notice that spreading sequences need to be used in boosting the estimation SNR. Different from the point-to-point scenario, though, is that there is more than one STA going to transmit training sequences at the same time. It would thus be beneficial to allocate orthogonal (or nearly orthogonal) spreading sequences for each STA during each antenna training step.

In Algorithm 4, we illustrate the iterative antenna training protocol for a time division duplexing system, where the uplink and downlink channels are reciprocal thanks to calibrated RF. For simplicity, we assume that $N \geq M$. The iterative method works by repeating steps (1) and (2) a number of times, where the outcomes $\mathbf{v}, \mathbf{w}$ from (4.30), after normalization, are used in the next step (2) and the outcomes $\mathbf{u}_1, \mathbf{u}_2$ from (4.31), after normalization, are used in the next step (1) [16].

Algorithm 4 Iterative protocol for SDMA

---

0. *Initialization*

0.1. We start by picking a random pair of orthogonal, non-zero vectors $\mathbf{u}_1 \perp \mathbf{u}_2$.

1. *STA side antenna training*

1.1 Keep sending the same vector $\mathbf{u}_1$ for N consecutive time slots from the first RF chain of the AP, while each time slot is spread by a pseudo-random spreading sequence p_1.

1.2 Keep sending the same vector $\mathbf{u}_2$ over N time slots from the second RF chain of the AP, while each time slot is spread by a pseudo-random spreading sequence p_2, that is (near) orthogonal to p_1.

1.3 At STA1, use $\mathbf{I}_N$ as the receive beamforming matrix over N time slots, while despreading using p_1 is performed.

1.4 In the meantime, at STA2, use $\mathbf{I}_M$ as the receive beamforming matrix over M time slots, while despreading using p_2 is performed. Collecting the received samples, we arrive at

$$\mathbf{v} \leftarrow \mathbf{H}_1'(\mathbf{u}_1 \oplus \mathbf{u}_2) + \text{noise}$$
$$\mathbf{w} \leftarrow \mathbf{H}_2'(\mathbf{u}_1 \oplus \mathbf{u}_2) + \text{noise} \tag{4.30}$$

where $\oplus$ indicates that $\mathbf{u}_1$ is spread by sequence p_1 and $\mathbf{u}_2$ is spread by sequence p_2 before they are added up in the air, and that corresponding despreading is employed at each STA. As a result of (4.30), we arrive at interim beamforming vectors $\mathbf{v}, \mathbf{w}$, which will be used immediately in the next step.

2. AP side antenna training

2.1 Keep sending $\mathbf{v}$, after normalization, over K time slots from STA1 while each time slot is spread by a pseudo-random sequence q_1.

2.2 Keep sending $\mathbf{w}$, after normalization, over K time slots from STA2 while each time slot is spread by a pseudo-random sequence q_2, that is (near) orthogonal to q_1.

2.3 For the first RF chain of the AP, use $\mathbf{I}_K$ as the receive beamforming matrix over K time slots while despreading using q_1 is performed.

2.4 For the second RF chain of the AP, use $\mathbf{J}_K$ as the receive beamforming matrix over K time slots while despreading using q_2 is performed. Here $\mathbf{J}_K$ is chosen as an orthogonal matrix such that each column of $\mathbf{J}_K$ is orthogonal to the corresponding column of $\mathbf{I}_K$.

2.5 Rearranging the collected samples, we arrive at

$$\mathbf{u}_1 \leftarrow \mathbf{H}_1\mathbf{v} + \text{interference} + \text{noise}$$
$$\mathbf{u}_2 \leftarrow \mathbf{H}_2\mathbf{w} + \text{interference} + \text{noise} \tag{4.31}$$

The interim beamforming vectors $\mathbf{u}_1, \mathbf{u}_2$ are to be used in the step 1 of the next iteration.

In this section, we have focused on antenna training for high-rate data links where directional transmission/reception is desired. On the other hand, for control signaling and/or low-rate data packets, omnidirectional transmission is still required. Omnidirectional transmissions can be achieved in different ways. One possible solution is to use omnidirectional transmissions for control messages. For example, the 2.4/5 GHz unlicensed band may be used as an out-of-band control channel in delivering the control messages for 60 GHz wireless networks. Another possible solution is to use multiple directional transmissions of the same signal to mimic an omnidirectional transmission. This is best illustrated in the example of Figure 4.11(c), where the six directional transmissions together are equivalent to an omnidirectional transmission with the same range. Notice that the same signals need to be transmitted in the six different directions to form the omnidirectional equivalent. In such a case, all control packets that are broadcasting messages in nature need to be transmitted in six replicas over the six different directions. Such packets include request-to-send (RTS), clear-to-send (CTS) and acknowledged/not acknowledged (ACK/NAK). More discussion on medium access control design directive transmission can be found in Chapter 8.

4.4 Summary

We have briefly introduced the wireless (indoor) channel characteristics in 60 GHz, where we face significant link budget challenges due to severe path loss, oxygen absorption and penetration loss. Practical channel measurements found that the main propagation mechanisms are line-of-sight, first- and second-order reflection paths, and that lower-order MIMO techniques would be more appropriate than higher-order MIMO techniques. Antenna array beamforming is preferred for 60 GHz wireless due to its large antenna gain, small size and fast electronic steerability, while the latter is especially meaningful in solving the notorious NLOS obstruction problem in real time.

To solve the NLOS obstruction problem in real time, the transmit/receive antenna arrays need to be equipped with a smart logic that automatically detects loss of LOS path, automatically finds the best current path (possibly one of the reflective paths), and automatically routes the high-speed data packets via the newly selected path. For this reason, a fast, efficient antenna array training method becomes one of the cornerstones in enabling Gbps wireless in 60 GHz.

For fully adaptive antenna arrays, we introduce an iterative antenna training method in which the transmitter beamforming coefficients and receiver beamforming coefficients are alternatively updated to make the achievable beamforming gain as large as possible. For switched antenna arrays at both transmitter and receiver, we introduce a tree search antenna training method where we have taken advantage of the two-level beamforming codebooks assumed at both transmitter and receiver.

High-speed data links in 60 GHz would typically be directional. Spatial division multiple access can then be used to allow the wireless spectrum to be utilized more efficiently and is expected to be used in future 60 GHz wireless networks.

References

[1] Xia, P., Yong, S. K., Oh, J. and Ngo, C. (2008) Multi-stage antenna training millimeter wave communication systems. *IEEE Globecom Conference 2008*.
[2] Xia, P., Niu, H., Oh, J. and Ngo, C. (2008) Practical antenna training for in-vehicle millimeter wave communication systems. *IEEE Vehicular Technology Conference*, Fall.
[3] Xia, P., Yong, S. K., Oh, J. and Ngo, C. (2008) A practical SDMA protocol for 60 GHz millimeter wave communications. *IEEE Asilomar Conference 2008*.

[4] Yong, S. K. (2007) TG3c Channel Modeling Sub-committee Final Report. IEEE 15-07-0584-01-003c, March.

[5] Shiraki, Y., Ohyama, T., Nakabayashi, S., Tokuda, K., Kato, A., Fujise, M. and Horimatsu, T. (2002) Experimental system of 60 GHz millimeter wave band inter-vehicle communications based CSMA method. *IEEE Intelligent Vehicle Symposium*, vol. 2, pp. 17–21.

[6] Yamamoto, A., Ogawa, K., Horimatsu, T., Kato, A. and Fujise, M. (2008) Path-loss prediction models for inter-vehicle communication at 60 GHz. *IEEE Transactions on Vehicular Technology*, **57**(1), 65–78.

[7] Maltsev, A. et al. 60 GHz WLAN experimental investigations. IEEE 802.11-08-1044r0.

[8] Tse, D. and Viswanath, P. (2005) *Fundamentals of Wireless Communication*. Cambridge: Cambridge University Press.

[9] Xu, H., Kukshya, V. and Rappaport, T. (2002) Spatial and temporal characteristics of 60-GHz indoor channels. *IEEE Journal on Selected Areas in Communications*, **20**(3), 620–630.

[10] Doan, C. H., Emami, S., Sobel, D. A., Niknejad, A. M. and Brodersen, R. W. (2004) Design considerations for 60 GHz CMOS radios. *IEEE Communications Magazine*, **42**(12), 132–140.

[11] Telatar, E. (1999) Capacity of multi-antenna Gaussian channels. *European Transactions on Telecommunications*, **10**(6), 585–596.

[12] Golub, G. H. and Van Loan, C. F. (1990) *Matrix Computations*. Baltimore, MD: Johns Hopkins University Press.

[13] Yong, S. K. (2007) TG3c Channel Modeling Sub-committee Final Report. IEEE 15-07-0584-01-003c, March.

[14] Bolcskei, H., Borgmann, M. and Paulraj, A. J. (2003) Impact of the propagation environment on the performance of space-frequency coded MIMO-OFDM. *IEEE Journal on Selected Areas in Communications*, **21**(3), 427–439.

[15] Cooper, M. and Goldburg, M. Intelligent antennas: spatial division multiple access. Arraycomm White Paper.

[16] Millimeter Wave Multi-Resolution Beamforming (2008) IEEE 802.15-08-0182.

5

Baseband Modulation

Pengfei Xia and André Bourdoux

5.1 Introduction

The choice of baseband modulation scheme for 60 GHz wireless depends on many factors such as the channel characteristics, deployment scenarios, antenna configurations, applications, spectrum efficiency and circuit limitations.[1] In general, three types of modulations are suitable candidates for 60 GHz communications, namely minimum shift keying (MSK), single carrier (SC) block transmissions and multi-carrier (MC) transmissions. A SC would be used to carry the MSK modulated information symbols as well. In this sense, MSK is itself a SC modulation scheme. We note that the term "single carrier" modulation in this chapter mainly refers to the SC block transmissions with general non-constant-envelope constellations, for example M-QAM. For MC transmissions, we focus on orthogonal frequency division multiplexing (OFDM) in this book.

MSK is a promising modulation scheme within the family of constant-envelope modulation. Its constant modulus with no peak-to-average power ratio (PAPR) enables high-efficiency operation of transmit power amplifiers (PAs) as described in Section 3.2. This makes MSK a very appealing modulation choice for 60 GHz. MSK also enjoys a high receive power efficiency in terms of power used to receive each bit. However, MSK suffers from relatively low throughput, especially in high signal-to-noise-ratio (SNR) regions, which limits its usage to the lower data rate range of multi-gigabit transmissions.

---

[1]The author would like to acknowledge Samsung Information Systems America, San Jose, CA, where the work for this chapter was done.

60 GHz Technology for Gbps WLAN and WPAN: From Theory to Practice
Edited by Su-Khiong (SK) Yong, Pengfei Xia and Alberto Valdes Garcia
© 2011 John Wiley & Sons, Ltd

For high-throughput systems, SC block transmissions (SCBTs) and OFDM modulations are the preferred choices due to the higher bandwidth efficiency, though at the cost of higher implementation complexity. Conceptually, OFDM transmits multiple carriers in parallel, with each occupying a narrow band, while SC modulation transmits a SC modulated at a high symbol rate.

The primary advantage of OFDM modulation relative to its SC counterpart is its reduced channel equalization complexity in the frequency domain, thanks to the sub-carrier orthogonality. On the other hand, for SCBT, a tap-delay line (time domain) equalizer is generally needed to equalize the channel, leading to a much more complicated receiver than that for OFDM transmissions [1]. The complexity of the time-domain equalizer becomes more critical in the case of Gbps high-speed transmission. In order to overcome this problem, SCBT with frequency-domain equalization (SC-FDE) has been proposed in [2–4] and the references therein.

Figure 5.1 shows system diagrams for OFDM and SC-FDE, which share highly similar system operations. For OFDM, fast Fourier transform (FFT) and inverse FFT operations reside in the receiver and transmitter, respectively, and the linear MMSE channel equalization is applied after FFT operation. On the other hand, for SC-FDE, both inverse FFT and FFT operations are carried out at the receiver side and the linear MMSE channel equalization is applied after FFT. Hence, the SC-FDE receiver has approximately double the complexity of the OFDM receiver but a simpler SC-FDE transmitter.

For the same constellation points, SC modulation has lower PAPR value than OFDM modulation. This places a more stringent requirement on PAs for OFDM transmission and thus lowers the overall power efficiency of the system. Nevertheless, it is realized that for SC to achieve high spectral efficiency, the use of high-order constellations (e.g. 16/64-QAM) is almost inevitable, which leads to a PAPR increase for SC as well. In such cases, the PAPR difference between OFDM and SC becomes insignificant.

OFDM suffers a severe problem with PAPR that decreases the efficiency of the PA. This problem is more critical in 60 GHz than for implementations below 6 GHz. Table 5.1 shows typical parameters for PA implementations below 6 GHz and at 60 GHz [5]. Additional information on 60GHz PA performance is presented in Table 6.5.

For operation below <6 GHz, the values for $P_{1\,dB}$ and P_{sat} are very close, indicating that the PA operates near the maximum power limit within the linear region. Hence, no or minimal PA output backoff is required. In contrast, the 60 GHz PA operates in a power-limited regime in order to maintain linearity. This requires large output backoff in case of OFDM modulation with high constellation points.

The performance of a modulation scheme also depends on the error control schemes such as error correction code itself and interleaving depth. In particular, uncoded OFDM modulation over wireless multi-path channels would suffer from

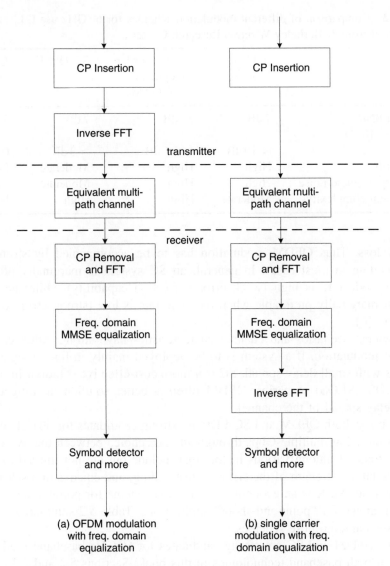

Figure 5.1 System diagrams for (a) OFDM and (b) SC-FDE.

Table 5.1 Comparison of typical parameters (dBm) for OFDM modulation in <6 GHz and 60 GHz [5]. Reproduced by permission of © 2009 IEEE

Parameter	<6 GHz	60 GHz
$P_{1\,dB}$	25.5	6.7
P_{sat}	26.5	9.8

Table 5.2 Comparison of different modulation schemes for 60 GHz use [7]. Reproduced from © Berkeley Wireless Research Center

Modulation	OFDM-QPSK	SC high-order modulation (16-QAM)	SC QPSK	Constant envelope (MSK)
Required SNR (BER = 10^{-3})	7 dB	12 dB	7 dB	7 dB
Tx PAR	≈ 10 dB	≈ 5.5 dB	≈ 3 dB	0 dB
PA linearity	High	High	Moderate	Low
Sensitivity to phase noise	High	High	Moderate	Low
Equalization complexity	Moderate	High	High	High

diversity loss. Thus OFDM modulation has to be accompanied by strong error control coding in most cases. In general, an SC system is marginally preferable when the code rate is high (weak error correction capability) while the OFDM system is marginally preferable when the code rate is low (strong error correction capability) [6].

Deployment scenarios and channel characteristics have strong influence on the choice of modulation. If a system is to be deployed mainly in line-of-sight (LOS) scenarios with small delay spread, SC is a more cost-effective solution. In contrast, in non-LOS (NLOS) scenarios, OFDM offers a better solution in mitigating the longer delay spread of the channel.

In summary, both OFDM and SC-FDE are strong candidates for 60 GHz communications targeting multiple-Gbps throughput. The choice between the two requires a careful tradeoff among various factors that include but are not limited to size of the constellation, strength of the error control coding, and equalization scheme. On the other hand, MSK is an economic modulation scheme for portable devices, and is a good choice for "point-and-shoot" applications. Table 5.2 summarizes the key comparisons among those schemes.

As both OFDM and SC are strong candidates for 60 GHz baseband modulation, we present both baseband technologies in this book. Sections 5.2 and 5.3 are dedicated to OFDM modulation (written by Pengfei Xia), while Sections 5.4 and 5.5 are dedicated to SC modulation (written by André Bourdoux).

5.2 OFDM Baseband Modulation

5.2.1 Principles of OFDM

Let X_n, $n = 0, 1, \ldots, N - 1$, be the N data symbols to be transmitted over the N carriers, where X_n can be represented in general as a complex point in a two-dimensional constellation, for example M-QAM or M-PSK. Let f_n be the carrier frequency for the nth carrier. The transmitted time-domain waveform can then be written as

$$x(t) = \sum_{n=0}^{N-1} X_n e^{j2\pi f_n t}. \tag{5.1}$$

A digital sampled version of the time-domain waveform is nothing more than

$$x(mT_s) = \sum_{n=0}^{N-1} X_n e^{j2\pi f_n mT_s}, \tag{5.2}$$

where $t = mT_s$ are the sampling points and T_s is the sampling period. Suppose that the N carriers are equally spaced in the frequency domain and that $f_n = nf_o$, $n = 0, 1, \ldots, N - 1$. In this case, the time-domain samplings become

$$x(mT_s) = \sum_{n=0}^{N-1} X_n e^{j2\pi n f_o mT_s}. \tag{5.3}$$

Furthermore, let $f_o = 1/NT_s$ be the minimum frequency separation to maintain sub-carrier orthogonality. The time-domain samples can be written as

$$x(mT_s) = \sum_{n=0}^{N-1} X_n e^{j2\pi mn/N}. \tag{5.4}$$

Clearly, the time-domain samples $\{x(0), x(T_s), \ldots, x((N - 1)T_s)\}$ are nothing more than the inverse discrete Fourier transform of the N data symbols $\{X_0, X_1, \ldots, X_{N-1}\}$, which can be implemented via FFT in a very efficient manner. Thus, inverse FFT is usually used at the OFDM transmitter. To recover the N data symbols from the time-domain samples received, it is natural to use FFT at the OFDM receiver.

For general MC transmissions over frequency selective multi-path fading channels, inter-symbol interference (ISI) and inter-carrier interference (ICI) arise. To keep the OFDM system from ISI and ICI, a cyclic prefix is usually inserted at the transmitter and then removed at the receiver. It can be shown that as long as the cyclic prefix length is longer than the channel delay spread, the linear convolution operation (thanks to the multi-tap channel response) is equivalent to a circular convolution operation [1]. A circular convolution of the transmitted waveform and the multi-tap channel in the time domain translates simply to a direct multiplication of the transmitted symbols and the channel frequency responses in the frequency domain. Henceforth, a low-complexity one-tap equalizer is sufficient. This is the most prominent benefit of OFDM communication systems.

The PAPR is defined as the ratio of the transmitted peak power versus the transmitted average power,

$$\frac{\max |x(t)|^2}{E[|x(t)|^2]}. \tag{5.5}$$

From Equation (5.1), the time-domain waveform $x(t)$ is essentially a summation of multiple independently modulated sub-carriers. Thus, there are some cases when all the sub-carriers are added up coherently, which leads to a PAPR as high as N. The high PAPR is problematic as it degrades the transmit power efficiency. To see this, we look at a generic PA input–output relationship in Figure 5.2 based on the well known Rapp model [8]. The Rapp model takes only the AM/AM effect into account.

Also illustrated is the maximum input power $P_{I,MAX}$, the average input power $P_{I,AVG}$, the maximum output power $P_{O,MAX}$ and the average output power $P_{O,AVG}$. It is noticed that the output power fully saturates at $P_{O,MAX}$ if the input power exceeds $P_{I,MAX}$. Thus to avoid saturation, the input power of the power amplifier has to be set at $P_{I,AVG}$ when a large PAPR signal is experienced. This leads to power backoff $10 \log_{10}(P_{O,MAX}/P_{O,AVG})$, which is excessive and undesirable. High PAPR is one major drawback of OFDM as compared with SC modulation schemes. Another major drawback of OFDM is its sensitivity to carrier frequency offset (CFO) and phase noise. Note that the phase noise issue is more pronounced at 60 GHz than at lower frequencies such as 5 GHz.

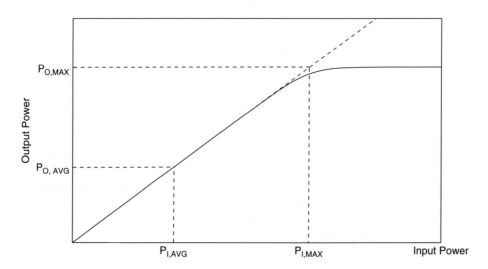

Figure 5.2 Illustration of the power amplifier Rapp model $g(A) = A/(1 + A^{2p})^{\frac{1}{2p}}$, $p = 10$.

Although the high PAPR and the CFO sensitivity issues appear to deter the application of OFDM, we have seen OFDM technology accepted and used in more and more standards and applications, such as digital audio broadcasting (DAB), digital video broadcasting (DVB), wireless LAN IEEE 802.11, IEEE 802.16 (WiMAX) and 3GPP LTE. Thus it would be reasonable to argue that those issues are problematic but not catastrophic, if proper care is taken to address them.

5.2.2   OFDM Design Considerations

Figure 5.3 roughly illustrates the typical time-domain waveforms of an OFDM system and an SC system, where the symbol period of the former is generally much larger than that of the latter. Conceptually speaking, timing synchronization for SC systems is more important because timing synchronization is more critical to handling ISI when the symbol period is small.

Figure 5.4 roughly illustrates the typical frequency content of an OFDM system and am SC system, where multiple orthogonal sub-carriers are seen in the former and only a single sub-carrier is seen in the latter. Differently, we expect more stringent requirements on frequency synchronization for OFDM systems because frequency synchronization is critical to maintaining sub-carrier orthogonality and handling CFO. In particular, bit error rate (BER) sensitivity to the CFO is analyzed for both OFDM and SC systems in [9], where we may approximate the degradation

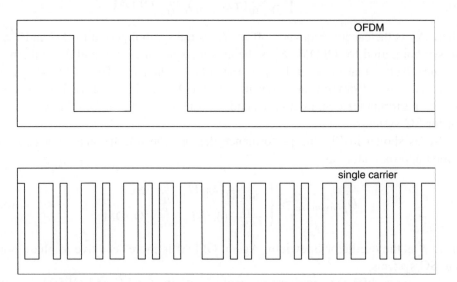

Figure 5.3   Time-domain illustration of OFDM and SC systems.

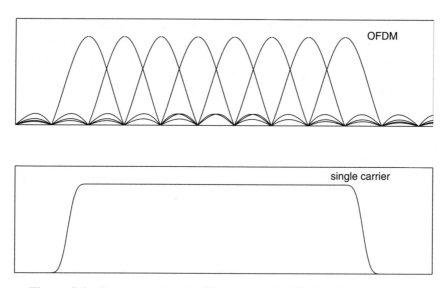

Figure 5.4   Frequency-domain illustration of OFDM and SC systems.

due to CFO as

$$
D_{CFO} \approx \begin{cases} \frac{10}{3\log 10}(\pi\Delta_f T_{sc})^2 & \text{SC} \\ \frac{10}{3\log 10}(\pi\Delta_f T_{mc})^2 \frac{E_s}{N_0} & \text{OFDM} \end{cases} \tag{5.6}
$$

where Δ_f is the carrier frequency offset, T_{sc} is the symbol period for SC and T_{mc} is the symbol period for OFDM, E_s is the average symbol energy and N_0 is the noise variance. Notice that in general T_{mc} is much larger than T_{sc}. Thus, OFDM systems are much (several orders) more sensitive to CFO than SC systems. It is observed that the performance degradation depends on the SNR ratio in OFDM systems but not for SC systems.

Also as shown in [9], the performance degradation due to phase noise can be written approximately as

$$
D_{phasenoise} \approx \begin{cases} \frac{1}{6\log 10}(4\pi\beta T_{sc})\frac{E_s}{N_0} & \text{SC} \\ \frac{11}{6\log 10}(4\pi\beta T_{mc})\frac{E_s}{N_0} & \text{OFDM} \end{cases} \tag{5.7}
$$

where β is the oscillator linewidth. Again, OFDM is more sensitive to phase noise than SC systems.

As we have different requirements and sensitivity for SC and OFDM systems, the design criteria for OFDM systems are different from those of SC systems. In the following, we discuss several key design considerations for OFDM systems.

- Number of sub-carriers (N_s)
  - Increasing N_s lengthens the OFDM symbol, reduces ISI and thus simplifies equalizer designs.
  - However, increasing N_s also increases (frequency) synchronization requirements and increases PAPR.
- OFDM symbol period (T_s)
  - In general, it is desirable to have $T_s \ll T_{coh}$. This condition ensures that the entire OFDM symbol period is less than the channel's coherence time T_{coh}, which is necessary to maintain sub-carrier orthogonality within the OFDM symbol.
  - The OFDM symbol period cannot be too small. As indicated below, OFDM cyclic prefix length needs to be taken into account to keep the transmit power efficiency from being too low.
- OFDM cyclic prefix (CP) length L_{cp}
  - CP Length needs to be larger than the channel delay spread to guarantee ISI-free reception.
  - Usually the channel delay spread is described in terms of its root-mean-square channel delay spread σ_h. A rule of thumb is to maintain $L_{cp} > 5\sigma_h$. It is noted that the transmit/receive filter responses should be taken into account when calculating the overall channel delay spread. Furthermore, a part of the guard interval needs to be reserved for synchronization margins due to imperfect acquisition time, and clock offset between transmitter and receiver.
  - Note that the transmit power is wasted during the CP period. In general, it is desirable to have the transmit power efficiency $(N_s + L_{cp})/N_s \leq 1$ dB.
- Sub-carrier bandwidth
  - Sub-carrier bandwidth should be small enough such that within each sub-carrier, the channel can be deemed to be frequency flat.
  - On the other hand, the sub-carrier bandwidth should be kept much larger than the channel Doppler spread f_d so that frequency synchronization task is doable.

As we can see, choice of OFDM system parameters is a tradeoff among various conflicting requirements. Outlined in the following is a simple design routine that works for most OFDM systems [1].

1. Set cyclic prefix length L_{cp} to be 2–4 times the delay spread.
2. Set OFDM symbol period T_s to be 5–6 times the cyclic prefix length.
3. Obtain the OFDM sub-carrier spacing as $1/(T_s - L_{cp})$.
4. Compute the number of sub-carriers required by dividing the system bandwidth by sub-carrier spacing. Alternatively, compute the number of sub-carriers by dividing the required data rate by the data rate per sub-carrier. The data rate

Table 5.3 OFDM system parameters for popular wireless systems [10]. Reproduced by permission of © 2007 John Wiley & Sons, Inc.

	DAB	IEEE 802.11	IEEE 802.16	3GPP LTE
Carrier freq (GHz)	1.5	2.5/5.8	2-11	2
Sample freq (MHz)	2	20	32	15.36
Bandwidth (MHz)	1.5	20	28	10
FFT size	1024	64	256	1024
Used sub-carriers	768	52	200	601
Guardband ratio	0.25	0.1875	0.2185	0.41
Carrier space (KHz)	2	312.5	125	15
FFT period (μs)	500	3.2	8	66.7
Guard interval (μs)	123	0.8	2/1/0.5/0.25	4.67/16.67
Constellation	DQPSK	PSK, QAM	PSK, QAM	PSK, QAM

per sub-carrier is defined by the M-QAM/M-PSK constellations, error control coding rate as well as symbol rate.

Table 5.3 lists design parameters for some popular OFDM systems, including digital audio broadcasting (DAB), IEEE 802.11 Wi-Fi, IEEE 802.16 WiMAX and 3GPP LTE.

The above design considerations are for general OFDM systems. For 60 GHz wireless systems based on OFDM, we do face some unique challenges. For example, phase noise in 60 GHz is excessive and could be as high as −85 dBc/Hz [11], compared with −110 to −120 dBc/Hz phase noise for general OFDM systems with carrier frequency below 6 GHz. Thus even more care should be exercised in dealing with frequency synchronization and tracking.

5.3 Case Study: IEEE 802.15.3c Audio Video OFDM

We next use the IEEE 802.15.3c audio video (AV) OFDM physical layer (PHY) mode [12] as a case study of OFDM modulation in 60 GHz. To support both high-rate and low-rate AV communications, two sub-modes are implemented. One is the so-called high-rate PHY (HRP) and the other is the so-called low-rate PHY. We here discuss the HRP mode only with a reference implementation of the HRP baseband illustrated in Figure 5.5.

5.3.1 Uncompressed Video Communications

A salient feature of the AV OFDM PHY mode in IEEE 802.15.3c is its capability to efficiently support uncompressed high-definition AV communications. Essentially,

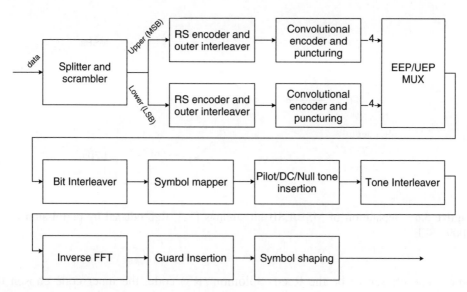

Figure 5.5 Block diagram of the IEEE 802.15.3c AV OFDM reference implementation [12]. Reproduced by permission of © 2009 IEEE.

it is designed to be the wireless high-definition multimedia interface (HDMI) which has become a standard interface in modern consumer communications.

A unique characteristic with uncompressed AV communications is that different bits are not created equal from the user's point of view. For example, a white pixel can be represented by an RGB 3-tuple $(R : 255, G : 255, B : 255)$ or in binary format $(R : 11111111_2, G : 11111111_2, B : 11111111_2)$. Apparently for each color from a recipient's point of view, the first '1' bit carries the most weight, the bits in the middle carry decreasing weights, while the last '1' carries the least weight. In other words, bit error with the first bit is most easily noticeable by a viewer, while bit error with the last bit is least easily noticeable. It is thus reasonable to apply stronger error protection to the more important bits.

For this reason, the incoming bit streams are split in AV OFDM into two classes at the very beginning (Figure 5.6). In particular, the incoming bit streams consist of bit streams 1–8 with bit streams 1–4 known as most significant bits (MSBs) and bit streams 5–8 known as least significant bits (LSBs). With bit streams differentiated as such, different error protection may be applied later. Random bit scrambling is then performed to avoid long runs of 0 s and 1 s.

5.3.2 Equal and Unequal Error Protection

As mentioned earlier, error control coding is indispensable for OFDM modulated systems. For MSBs/LSBs in AV OFDM, a concatenated code is used, with the

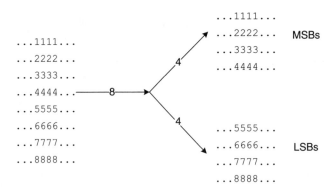

Figure 5.6 Separation of MSB/LSB bit streams [12]. Reproduced by permission of © 2009 IEEE.

outer code chosen to be the Reed–Solomon (RS) code, the inner code chosen to to be the standard convolutional code, and with an outer interleaver in the middle. Such a code concatenation has seen many applications, for example deep space communications.

In particular, a $(224, 216, t = 4)$ RS code is used as the outer code, which may be obtained by shortening a standard $(255, 247, t = 4)$ RS code. The code is able to correct $t = 4$ error symbols, with each symbol being an 8-bit element in the $GF(2^8)$ Galois field whose field generator polynomial is given by

$$g_{field}(x) = x^8 + x^4 + x^3 + x^2 + 1. \tag{5.8}$$

The RS code generator polynomial can be written as

$$g_{RS}(x) = (x + \lambda)(x + \lambda^2)(x + \lambda^3)(x + \lambda^4)(x + \lambda^5)(x + \lambda^6)(x + \lambda^7)(x + \lambda^8) \tag{5.9}$$

with $\lambda = 0x02$.

The rate 1/3 convolutional code is used as the inner code, with constraint length $K = 7$, and generator polynomial $g_X = 133_8$, $g_Y = 171_8$, $g_Z = 165_8$. A reference implementation of the encoder is illustrated in Figure 5.7.

At the receiver, the Viterbi decoder needs to decode the bit streams at a throughput of the order of several Gbps, which is quite a challenge using a single decoder with the current technology. For this reason, code parallelization is used so that the total bit streams are encoded separately by a bank of eight convolutional encoders. At the receiver, the received signal may be decoded separately by a bank of eight (hard/soft) Viterbi decoders. In the following, the eight parallel convolutional encoders are labeled A–H, with A–D for MSBs and E–H for LSBs.

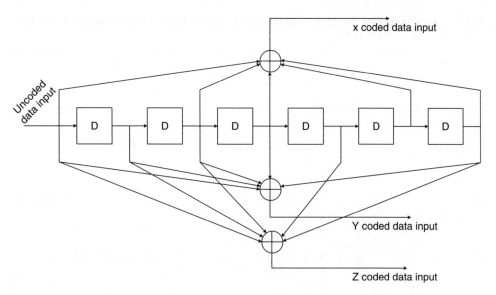

Figure 5.7 A reference implementation of the rate 1/3 convolutional code [12]. Reproduced by permission of © 2009 IEEE.

As discussed earlier, it is beneficial to apply unequal error protection (UEP) to video transmissions by applying different error protection to the MSBs and LSBs. One way of implementing UEP in AV OFDM is by using different coding rates for MSBs and LSBs. This is made possible by convolutional code puncturing. In Table 5.4, puncturing pattern '1' means transmitting the bit while puncturing pattern '0' means skipping the corresponding bit. Code rates 1/3, 1/2, 2/3 may be used for bit streams in the equal error protection (EEP) mode, rates 4/7, 4/5 may be used for MSBs and LSBs respectively in the UEP mode. In general, MSBs should be protected by a lower rate code (stronger code) than LSBs.

It is noted that other options remain to implement UEP for MSBs and LSBs in the physical layer. For example, in an OFDM modulated system, MSBs and LSBs may be allocated to different sub-carriers and MSB-sub-carriers may be allocated more power than LSB sub-carriers so that stronger error protection is achieved for MSBs. Another possibility would be to distribute MSBs and LSBs respectively to the I branch and Q branch of the signal constellation and have larger constellation distance over the I branch than the Q branch. Stronger error protection is thus achieved for MSBs as well.

Overall, the UEP via modulation approach is very flexible to allow different level of UEP since the I/Q distance ratio can be finely adjusted. The UEP via power loading approach (for OFDM/MC modulation systems only) is also flexible since the power loading across different sub-carriers can be finely adjusted as well. These

Table 5.4 Puncturing patterns to achieve different coding rates [12]. Reproduced by permission of © 2009 IEEE

Code rate	Puncturing pattern	Transmitted sequence
1/3	X: 1 Y: 1 Z: 1	$X1, Y1, Z1$
1/2	X: 1 Y: 1 Z: 0	$X1, Y1$
4/7	X: 1 1 1 1 Y: 1 0 1 1 Z: 0 0 0 0	$X1, Y1, X2, X3, Y3, X4, Y4$
4/5	X: 1 1 1 1 Y: 1 0 0 0 Z: 0 0 0 0	$X1, Y1, X2, X3, X4$
2/3	X: 1 1 Y: 1 0 Z: 0 0	$X1, Y1, X2$

two approaches lead to potentially larger PAPR because of the power imbalance. On the other hand, the UEP via coding approach has no impact on the system PAPR and is thus desirable from this perspective.

5.3.3 Bit Interleaving and Multiplexing

The encoded bits from the eight parallel encoders need be multiplexed together to be mapped onto complex constellations. In AV OFDM, QPSK and 16-QAM constellations are used. For 16-QAM, every four bits are mapped to a constellation point – two bits to the I branch and the other two bits to the Q branch. It is noted that for the two bits allocated to the I branch, one bit is better protected and will be labeled as the better protected bit (BPB) in the following, while the other bit will be labeled as the worse protected bit (WPB). The same situation occurs for the Q branch.

The aim of bit interleaving, after multiplexing, is that the encoded bits from each of the eight convolutional encoders get a more or less equal chance of being BPBs and WPBs. If the encoded bits from one particular convolutional encoder all ended up in the WPB positions, the performance degradation would be non-negligible. In particular, it would be desirable to have encoded bits from each encoder to alternate in being BPBs and WPBs. In other words, if the current bit from encoder A is mapped to be a BPB, it would be desirable to have the previous/next bit from encoder A mapped to be a WPB. In this way, the interleaved bits would be well distributed to take advantage of the error correction capability provided by the convolutional codes. Note that the convolutional codes and the Viterbi decoder are known to handle random errors well, but not bursty errors.

Furthermore, when the encoded and interleaved bits are mapped onto the constellation symbols, it is desirable to have the bits from one encoder mapped onto multiple different constellation symbols. This would take advantage of the underlying channel diversity when different constellation symbols experience different, and independent, fading channels. In summary, we have the following design criteria for designing bit interleavers.

1. Alternate the encoded bits from each encoder (e.g. A1–A7 for encoder A) to be BPBs and WPBs.
2. Avoid mapping the encoded bits from the same encoder onto the same constellation symbol.

Bit interleaving in AV OFDM is achieved by a block interleaver of size 48. In the EEP mode (convolutional code rate 2/3 for each of the eight encoders), it accepts six encoded bits from each convolutional encoder. In the following we use A1–A6 to represent the six bits from encoder A, and similarly for encoders B–H. The bit interleaving is done such that the bits are ordered as illustrated in Figure 5.8. In particular, the bits are ordered from left to right and from top to bottom. For each column, the four bits are mapped to be BPB in the I branch, WPB in the I branch, BPB in the Q branch and WPB in the Q branch. It can be verified that the encoded bits from each encoder alternate in being BPBs and WPBs.

Column-wise read out, from top to bottom and from left to right $\longrightarrow$

I–BPB	A1	E5	D3	C1	G5	F3	E1	A5	H3	G1	C5	B3
I–WPB	B2	F6	E4	D2	H6	G4	F2	B6	A4	H2	D6	C4
Q–BPB	C3	B1	F5	E3	D1	H5	G3	F1	B5	A3	H1	D5
Q–WPB	D4	C2	G6	F4	E2	A6	H4	G2	C6	B4	A2	E6

Figure 5.8 Ordering at the bit interleaver output: EEP [12]. Reproduced by permission of © 2009 IEEE.

In the UEP mode (convolutional code rate 4/7 for encoders A–D and 4/5 for encoders E–H), it accepts seven encoded MSB bits from encoders A–D and five encoded LSB bits from encoders E–H. In the following we use A1–A7 to represent the seven bits from encoder A, and similarly for encoders B–D; E1–E5 to represent the five bits from encoder E, and similarly for encoders F–H. The bit interleaving is done such that the bits are ordered as illustrated in Figure 5.8. In particular, the bits are ordered from left to right and from top to bottom. For each column, the four bits are mapped to be BPB in the I branch, WPB in the I branch, BPB in the Q branch and WPB in the Q branch.

It is noted that for encoder A in the EEP mode, the encoded bits A1–A6 alternate as BPB, WPB, BPB, WPB, BPB, WPB; in the UEP mode, the encoded bits A1–A7 alternate as BPB, WPB, BPB, WPB, BPB, WPB and BPB. Thus design criterion 1 is met. Furthermore, for encoder A in the EEP mode, the encoded bits A1–A6 are mapped onto the 1st, 11th, 10th, 9th, 8th, and 6th 16-QAM symbols;[2] in the UEP mode, the encoded bits A1–A7 are mapped onto the 1st, 11th, 10th, 9th, 8th, 6th and 2nd 16-QAM symbols. Thus design criterion 2 is met.

It is noted that although the bit orderings appear different for the EEP and UEP modes, they share the same digital logic in implementation. To see this, we enumerate the encoded bits from the eight encoders in the natural order. In particular, for the EEP mode, bits 0–47 represents the encoded bits A1–A6, B1–B6, . . . , H1–H6 in order. For the UEP mode, bits 0–47 represent the encoded bits A1–A7, . . . , D1–D7, E1–E5, . . . , H1–H5 in order. Let $x = 0, \ldots, 47$ be the bit index at the input of the bit interleaver and $y = 0, \ldots, 47$ be the bit index at the output of the bit interleaver. It can be shown that for both EEP and UEP mode, the following relationship between x and y holds for Figures 5.8 and 5.9:

$$y = \mathrm{mod}\left(6 \times \left\lfloor \frac{x}{6} \right\rfloor - 5 \times \mathrm{mod}(x,6), 48\right) \tag{5.10}$$

where mod is the standard integer modular operation and $\lfloor \cdot \rfloor$ is the standard floor operation.

5.3.4 AV OFDM Modulation

Size 512 FFT is used for OFDM modulation in the AV OFDM PHY mode. The 512 HRP sub-carriers are arranged as per Table 5.5, where data sub-carriers carry complex data symbols, pilot sub-carriers carry pilot symbols, and DC and null sub-carriers carry null symbols. The complex symbols are interleaved once again before the inverse FFT (IFFT) operation by a bit-reversal tone interleaver. The tone interleaver is intended to scramble the positions of the originally adjacent symbols

---

[2]One 16-QAM symbol is a column in Figures 5.8 and 5.9.

————————— Column-wise read out, from top to bottom and from left to right —————————→

I–BPB	A1	E1	C7	B6	G3	E5	D4	A5	H2	F4	C3	B2
I–WPB	B1	F3	D7	C6	H5	G2	E4	B5	A4	H1	D3	C2
Q–BPB	C1	A7	F2	D6	C5	H4	G1	E3	B4	A3	G5	D2
Q–WPB	D1	B7	G4	F1	D5	A6	H3	F5	C4	B3	A2	E2

Figure 5.9 Ordering at the bit interleaver output: UEP [12]. Reproduced by permission of © 2009 IEEE.

Table 5.5 AV OFDM HRP sub-carrier arrangement [12]. Reproduced by permission of © 2009 IEEE

Sub-carrier type	Sub-carrier number k
Null sub-carrier	$k = [-256 : 1 : -178] \cup [178 : 1 : 255]$
Pilot sub-carrier	for sym = 0: 1: $N_{symbol} - 1$
	{
	$\quad k = [-177 + \mathrm{mod}(3 \times \mathrm{sym}, 22) : 22 : 177]$
	$\quad k \neq -1$, or 0 or 1
	}
DC sub-carrier	$k = [-1, 0, 1]$
Data sub-carrier	others

onto sub-carriers far apart so that they see relatively independent channel fadings from each other. In particular, let $0 \leq k \leq 511$ be the symbol index before the tone interleaver and $0 \leq \ell \leq 511$ be the symbol index after the tone interleaver. The bit-reversal tone interleaver dictates that the input symbol at position k,

$$k = \sum_{i=0}^{8} a_i 2^i, \tag{5.11}$$

be shuffled to the position ℓ,

$$\ell = \sum_{i=0}^{8} a_{8-i} 2^i. \tag{5.12}$$

Notice that k's binary representation $\{a_8, a_7, a_6, a_5, a_4, a_3, a_2, a_1, a_0\}$ is the bit-reversal of ℓ's binary representation $\{a_0, a_1, a_2, a_3, a_4, a_5, a_6, a_7, a_8\}$. It is known that regular IFFT operation consists of a bit reversal operation followed by an FFT butterfly operator. Thus, the bit-reversal tone interleaver is able to shuffle the adjacent complex symbols while at the same time reducing the IFFT complexity.

Among the remaining sub-carriers, 16 of them are used as pilot sub-carriers to facilitate channel estimation. Normally the pilot sub-carriers are equally spaced in the frequency domain to allow for simple channel estimation and interpolation. It is noted that the distance between two adjacent pilot sub-carriers needs to be smaller than the channel coherence bandwidth to allow for accurate channel estimation. A traveling pilot is used such that positions of the pilot sub-carriers change slowly and steadily with time. Thanks to the traveling pilot, time-domain and frequency-domain interpolation may be used to improve the channel estimate quality at the receiver side.

Null sub-carriers are placed at the two boundaries $\{-256 : 1 : -178\}$ and $\{178 : 1 : 255\}$ to prevent significant signal leakage into adjacent frequency bands. The collection of all these sub-carriers is also known as the guard band. Reservation of this guard band helps reduce the out-of-band emissions and relieves the front-end filter requirements in the RF domain. The direct current (DC) sub-carriers $\{-1, 0, 1\}$ are also reserved as null sub-carriers to simplify digital-to-analog and analog-to-digital conversions.

A list of modulation and coding modes is given in Table 5.6 for AV OFDM. Among them, modes 2 and 4 provide a nominal data throughput of 3.807 Gbps and are intended to carry a 3 Gbps 1080 p (p for progressive) HDMI signal (1920 columns per frame $\times$ 1080 rows per frame $\times$ 60frames per second $\times$ 24 bits per pixel $\approx$ 3 Gbps); modes 1 and 3 provide a nominal data throughput of 1.904 Gbps

Table 5.6 List of AV OFDM HRP modes [12]. Reproduced by permission of © 2009 IEEE

HRP mode index	Coding mode	Modulation	MSB code rate	LSB code rate	Data rate (Gbps)
0	EEP	QPSK	1/3	1/3	0.952
1	EEP	QPSK	2/3	2/3	1.904
2	EEP	16QAM	2/3	2/3	3.807
3	UEP	QPSK	4/7	4/5	1.904
4	UEP	16QAM	4/7	4/5	3.807
5	MSB-only retransmission	QPSK	1/3	n/a	0.952
6	MSB-only retransmission	QPSK	2/3	n/a	1.904

Modes 5 and 6 are for retransmission purposes where only the MSBs are retransmitted.

and are intended to carry a 1.5 Gbps 1080i (i for interlaced) HDMI signal. The optional modes 5 and 6 are intended for retransmissions, where only the MSBs are to be retransmitted while the bit errors with LSBs are ignored.

5.4   SC with Frequency-Domain Equalization

This section is concerned with SC transmissions with frequency-domain equalization at 60 GHz. As a case study, it describes the SC air interface selected by the IEEE 802.15.3c standard for wireless communications in the 60 GHz frequency band. It also provides a detailed description of the signal processing functionality required to support SC transmission for both time-domain and frequency-domain processing.

5.4.1   Introduction

SC modulation has been selected as one of the prominent PHY modes for IEEE 802.15.3c because it benefits from several very interesting properties, as shown in the following.

Low PAPR. The sensitivity of a signal to PA nonlinearity is closely linked to the characteristics of the signal envelope. Constant-envelope signals such as CPM and unfiltered M-PSK signals are the least sensitive. OFDM and pulse-shaped SC signals are not, however, constant-envelope and may suffer from PA nonlinearity. Figure 5.10 shows the complementary cumulative distribution function (CCDF) of the PAPR for OFDM and SC transmissions [13] and demonstrates that the PAPR of OFDM transmissions is quite high and is not dependent on the transmit filter nor on the constellation size. In contrast, the PAPR of SC transmissions is lower and does depend on the excess bandwidth of the pulse shaping filter and on the constellation. The lower the excess bandwidth, the higher the PAPR is. Hence, for SC-FDE, the PAPR can be traded off against spectral efficiency in a different manner for each constellation.

Low-complexity equalization. For SC transmissions over multi-path channels, time-domain equalizers are very complex to implement. The equalizer length depends on the channel length, which is not known in advance. In the SC case it is possible to go to the frequency domain in the receiver, perform the low-complexity single-tap equalization in the frequency domain just as for OFDM and then go back to the time domain. This implies the use of an FFT and an IFFT at the receiver but no IFFT is needed at the transmitter. The transmitter just needs to insert a cyclic prefix (see Section 5.5.2).

Possible tradeoffs on equalization complexity and transmission overhead. With SC transmission, the combinations of TX and RX processing indicated in Table 5.7

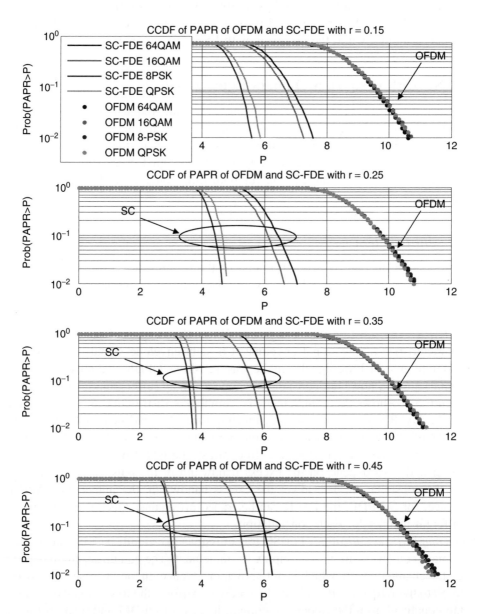

Figure 5.10 Complementary cumulative density function of the PAPR of SC and OFDM for different rolloff factors of the pulse shaping filter. The PAPR of SC varies with both the constellation order and rolloff factor, whereas the PAPR of OFDM is not affected by the constellation [13]. Reproduced by permission © 2008 IEEE.

Table 5.7 Possible uses of the SC and SC-FDE scheme according to channel conditions

Transmitter	FFT and IFFT receiver	Single-tap equalizer	Complexity and overhead
Yes	Yes	FDE	Higher complexity, CP overhead
Yes	No	TDE	Lower complexity, CP overhead
No	No	TDE	Lower complexity, no CP overhead

are possible. This provides a certain flexibility that can be exploited at run-time to minimize the power consumption and optimize the throughput according to the channel conditions.

Easy combination with spreading. When the link budget is not sufficient to ensure reliable communication, SC modulation can be easily combined with spreading to improve the SNR at the receiver. In this case, if the physical bandwidth has to be kept constant, the bit rate must be scaled by the inverse of the spreading factor.

Range of device classes and common mode. SC modulation can be used with different kinds of constellations such as BPSK, QPSK, 8-PSK and 16-QAM. Special – though simple – techniques such as $\pi/2$-BPSK or GMSK can also be used to reduce the sensitivity to PA nonlinearity. Different coding schemes can be used, with varying level of error protection. By designing transceivers for only a subset of the possible modulation and coding schemes (MCS), it is possible to define device classes with different complexity and power consumption. Furthermore, a simple, robust MCS – with very low complexity – can be defined to ensure compatibility with other devices that use a modulation other than SC.

5.4.2 Case Study: IEEE 802.15.3c SC PHY

The IEEE802.15.3c standard defines the PHY and MAC for short-range unlicensed communications, providing high-rate wireless personal area network (PAN) transport for bulk data transfer and multimedia streaming. It defines three different PHYs: one SC PHY and two OFDM PHYs. We give a brief summary of the SC PHY specification.[3]

The available bandwidth at 60 GHz varies over different areas of the world. The overlap of all frequency allocations covers the range 57–66 GHz. In order to be compatible with the frequency allocations in most areas, the IEEE802.15.3c standard has defined its operation in four different bands, as described in Table 5.8.

---

[3] Portions reprinted from [12]. Reproduced by permission of © 2009 IEEE.

Table 5.8   Channelization at 60 GHz

Channel number	Start frequency (GHz)	Center frequency (GHz)	Stop frequency (GHz)
1	57.240	58.320	59.400
2	59.400	60.480	61.560
3	61.560	62.640	63.720
4	63.720	64.800	65.880

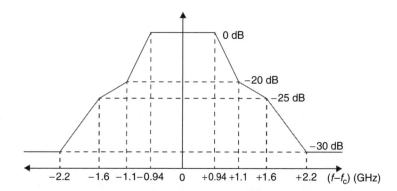

Figure 5.11   Transmit PSD mask [12]. Reproduced by permission of © 2009 IEEE.

This channelization is common to the SC and OFDM PHYs, and is also identical to the channelization in ECMA-387. It should be noted that the start and stop values in Table 5.8 are nominal values only. The occupied bandwidth in one channel is restricted by the transmit power spectral density (PSD) mask, which is as illustrated in Figure 5.11.

5.4.2.1   Frame Format and Preamble

The PHY frame consists of three parts: the PHY preamble, the frame header and the payload, as illustrated in Figure 5.12.

The PHY preamble is intended to support receiver processing related to AGC setting, antenna diversity selection, timing acquisition, frequency offset estimation, frame synchronization and channel estimation. The synchronization (SYNC) field is used for frame detection and uses a repetition of codes for higher robustness. The SYNC field consists of 14 repetitions of the Golay sequence a_{128} (see [12]). The start frame delimiter (SFD) field is used to establish frame timing as well as the header rate, either medium rate (MR) or high rate (HR). The MR header uses an SFD with

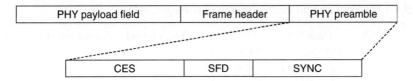

Figure 5.12   SC frame structure (transmit order is from right to left).

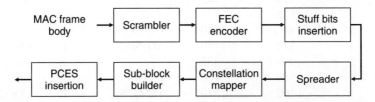

Figure 5.13   Payload field encoding and modulation [12]. Reproduced by permission of © 2009 IEEE.

$[+1\ -1\ +1\ -1]$ spread by $\mathbf{a}_{128}$ and the HR header uses an SFD with $[+1\ +1\ -1\ -1]$ spread by $\mathbf{a}_{128}$. The channel estimation (CES) field, used for channel estimation, consists of the sequence $[\mathbf{b}_{128}\ \mathbf{b}_{256}\ \mathbf{a}_{256}\ \mathbf{b}_{256}\ \mathbf{a}_{256}]$ where the rightmost sequence, $\mathbf{a}_{256}$, is first in time. The sequences $\mathbf{a}_{128}$, $\mathbf{a}_{256}$ and $\mathbf{b}_{256}$ are specified in [12].

The frame header is added after the PHY preamble. It conveys information in the PHY and MAC headers necessary for successfully decoding the frame. The frame header consists of a base frame header (containing the PHY header and the MAC header) followed by an optional frame header (containing the MAC sub-header). Note that the PHY header conveys the information about the MCS used the payload following the header. So it is essential that the header be received correctly.

The payload field is the last component of the frame. It contains the encoded and modulated information bits and is constructed as shown in Figure 5.13. The MAC frame body (i.e. the bits from the MAC) are first scrambled with bit sequence generated by means of a 16-stage linear feedback shift register (LFSR) as described in [12].

The scrambled bits are then encoded by one of the encoders in Table 5.9. Let K be the number of information bits and N be the length of the codeword. The forward error correction (FEC) code rate is then $r = N/K$. Stuff bits are then added to the end of the encoded MAC frame body if the number of encoded data bits is not an integer multiple of the length of the data portion in the sub-block. The stuff bits are set to zero and appended to the MAC frame body before scrambling. At the receiver, after decoding, the stuff bits are discarded. If spreading is applied, the encoded bits are spread with a spreading factor (SF) of 2, 4 or 64. For SFs 2 and 4, two or four consecutive output bits of the length-16 LFSR are used to spread the

Table 5.9 Modulation and coding scheme of the SC PHY

MCS class	MCS identifier	Data rate (Mbps) with pilot word length $= 0$	Data rate (Mbps) with pilot word length $= 64$	Modulation	Spreading factor	FEC type
Class 1	0	25.8 (CMS)	–		64	
	1	412	361		4	RS(255,239)
	2	825	722		2	
	3	1650 (MPR)	1440	$\pi/2$-BPSK/	1	
	4	1320	1160	(G)MSK	1	LDPC(672,504)
	5	440	385		2	(672,336)
	6	880	770		1	
Class 2	7	1760	1540	$\pi/2$-QPSK	1	LDPC(672,336)
	8	2640	2310	$\pi/2$-QPSK	1	LDPC(672,504)
	9	3080	2700	$\pi/2$-QPSK	1	LDPC(672,588)
	10	3290	2870	$\pi/2$-QPSK	1	LDPC(1440, 1344)
	11	3300	2890	$\pi/2$-QPSK	1	RS(255,239)
Class 3	12	3960	3470	$\pi/2$-8-PSK	1	LDPC(672,504)
	13	5280	4620	$\pi/2$-16QAM	1	LDPC(672,504)

information bits. For SF 64, the combination of the Golay sequences $\mathbf{a}_{64}$ and $\mathbf{b}_{64}$ and the length-16 LFSR is used.

5.4.2.2 Modulation Schemes

The encoded and spread bits are fed to the constellation mapper. Exact locations of the constellation points are illustrated in Figure 5.14, together with the mapping rules. Note that the 16-QAM constellation must be scaled by a factor $1/\sqrt{10}$ to yield the same average symbol energy as the other constellations.

The modulated symbols from the constellation mapper are split into blocks and the blocks are split into sub-blocks (Figure 5.15). A block contains 64 sub-blocks with the exception of the last block. A sub-block has length 512 and is formed by appending a pilot word to the data. The possible pilot word lengths are 0 and 64. For pilot word lengths 0 and 64, the length of the data is 512 and 448, respectively. Even-number blocks use pilot word sequences of type $\mathbf{a}_{64}$. Odd number blocks use pilot word sequences of type $\mathbf{b}_{64}$.

Furthermore, the output (denoted by c_i in Figure 5.15) of an LFSR shall be used to change the polarity of pilot word from one block to another. The LFSR used is the same as the length-16 LFSR, but is run at the appropriate rate: one LFSR

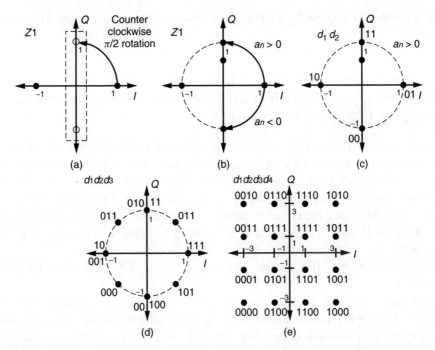

Figure 5.14 Constellation definition of the SC MCS: (a) $\pi/2$ BPSK; (b) pre-coded (G)MSK; (c) $\pi/2$ QPSK; (d) $\pi/2$ 8-PSK; (e) $\pi/2$ 16-QAM [12]. Reproduced by permission of © 2009 IEEE.

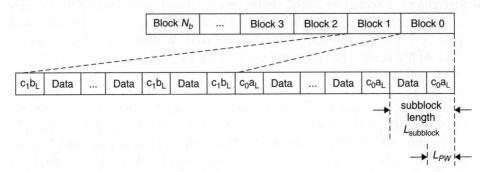

Figure 5.15 Block and sub-block formation for pilot word length 0 and 64 [12]. Reproduced by permission of © 2009 IEEE.

output per block. The last sub-block of the block is followed by a pilot word as well, to ensure that the cyclic extension is also correct at the sub-block boundaries. The pilot word will be modulated with $\pi/2$-BPSK irrespective of the constellation used for the data. An optional pilot word of length 9×128 symbols can be inserted between blocks to allow the channel to be reacquired.

Each symbol is rotated by a counter-clockwise $\pi/2$ sequence according to the formula:

$$x_n = j^n s_n, \quad n = 0, \ldots, N - 1,$$

where s_n and x_n are the symbol value before and after $\pi/2$ rotation respectively.

The MCSs supported by the IEEE802.15.3c standard are listed in Table 5.9. The following key features can be observed:

- The first MCS is intended for common mode signaling (CMS), which allows interoperability with non-SC devices. It is the most robust MCS with lowest rate, and is not intended for high-rate data transmissions. Rather, it is designed for control signaling and interoperability among all devices (OFDM/SC PHYs).
- The fourth MCS is called the mandatory PHY rate (MPR) and is defined to meet the PAR of IEEE802.15.3c, which is required to achieve a bit rate of at least 1 Gbps.
- The CMS and the MPR are the only mandatory MCSs. All other MCSs are optional.
- $\pi/2$ rotation between consecutive symbols is introduced to reduce the PAPR for BPSK. To simplify the hardware, it is also used with the other constellations.
- $\pi/2$ rotation also enables Gaussian minimum shift keying (GMSK) modulation by differentially encoding the bit streams. GMSK is known to have a PAPR of unity and be insensitive to PA nonlinearity. An analog modulator implementation is presented in Chapter 6.
- Spreading is foreseen for BPSK modes only because it provides an SNR increase (at the despreader output) at the cost of a bit-rate decrease. Hence, spreading with higher rate modulations (QPSK, etc.) would provide modes with similar rates to other BPSK modes. For example, QPSK with an SF of 4 has the same rate as BPSK with an SF of 2.
- For all MCSs – except for the first one – a pilot word of length 64 can be added. This enables the frequency-domain equalization (see Section 5.5.2). For MCS 0 (i.e. the CMS), which uses a very high SF of 64, the despreader can partially compensate for the channel multipath. Hence, this very robust and simple mode does not require very sophisticated processing. This is important since the CMS also has to be supported by devices using other PHYs.

5.5 SC Transceiver Design and System Aspects

5.5.1 Transmit and Receive Architecture

Several transceiver architectures can be used to transmit and receive complex modulations. We will concentrate on the receiver architecture, since the dual properties mostly also apply to the transmitter. As far as the receiver is concerned, the major functions are those of frequency down-conversion, quadrature demodulation, variable gain and sampling.

A first classification of receivers is super-heterodyne versus direct conversion:

- Super-heterodyne receivers make use of a down-conversion to baseband in at least two conversions (in other words, they make use of an intermediate frequency (IF) that is not centered on frequency 0). This usually comes with a tunable local oscillator that translates the desired RF frequency to a fixed IF frequency, where a highly selective filter rejects close-by interfering signals and selects only the desired channel. Variations of this receiver are the low-IF and sliding IF receiver. Chapter 6 describes the frequency planning for a sliding-IF 60 GHz system.
- Direct conversion receivers (also called zero-IF or homodyne receivers) use a local oscillator (LO) frequency centered exactly on the channel of interest. Hence, the desired channel is translated directly to baseband where the highly selective filtering can be implemented by means of sharp low-pass filters.

For this section, we will use the direct conversion architecture as the reference architecture for our analysis of the baseband processing. This is motivated by the following:

- It is the simplest architecture and facilitates implementation of 60 GHz transceivers.
- No highly selective bandpass, wideband IF filters are necessary.
- Because the bandwidth of the baseband I and Q signals is the half of the RF bandwidth, the analog-to-digital converter (ADC) can be run at a lower sampling rate.
- The direct conversion receiver uses analog quadrature generation. Hence, the non-idealities such as IQ imbalance and DC offset must be tackled. This makes our analysis also mostly valid for super-heterodyne receivers with analog quadrature demodulation.

A second classification of receivers is analog versus digital quadrature demodulation:

- In analog quadrature demodulation, the last down-conversion stage (which is also the unique down-conversion in direct conversion receivers) translates the channel spectrum to baseband by means of two analog LO signals in quadrature, yielding the in-phase and quadrature (I and Q) components of the complex baseband signal. Two ADCs are then used to digitize the signals, with a sample rate equal to or greater than the channel bandwidth. This architecture is known to suffer from IQ imbalance and DC offset.

- In digital quadrature demodulation, the desired spectrum is centered on a (usually low) IF and a single ADC is used to sample the real signal with a sample rate equal to or greater than twice the channel bandwidth. This is sometimes also referred to as bandpass sampling or IF sampling. Subsampling is possible in this case with some restrictions on the value of the sampling rate [14].

Analog quadrature generation is mostly used because it has the lesser ADC requirements in terms of sampling rate. The digital quadrature generation is difficult to use for 60 GHz transceivers because it requires a significant oversampling factor (3 or 4). Since analog quadrature generation is generally used in 60 GHz transceivers, IQ imbalance is usually a problem and ways to estimate and compensate for it are needed.

Figure 5.16 is a block diagram of the direct conversion architecture. In the receive path, bandpass filters are used at RF to pre-select a frequency range of interest (typically a few channels). An LNA provides gain and fixes the noise floor of

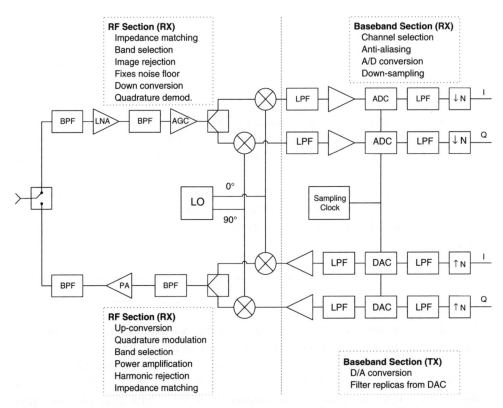

Figure 5.16   Direct conversion receiver and transmitter block diagram [15]. Reproduced by permission of © 2008 John Wiley & Sons, Ltd.

the system. An AGC amplifier brings the signal to a desired level for optimum analog-to-digital conversion at baseband. After the AGC amplifier, the signal is split into two equal components and directly down-converted to baseband with two LO signals in quadrature. The baseband signals are then low-pass filtered for two purposes: to avoid any aliasing and to reject adjacent channels. The next operation is analog-to-digital conversion. To meet the Nyquist sampling constraint, the sample rate F_S must be at least twice as high as half the signal bandwidth F_{BW}: $F_S > F_{BW}$. The operation of the transmit part is the dual of that of the receiver.

Figure 5.17 illustrates the main non-idealities present in direct-conversion transceivers. When the impact of these non-idealities is too high, they need to be estimated and compensated by the receiver signal processing. On the transmit side, we recognize the clipping and quantization of the digital-to-analog converter (DAC), the transmit IQ imbalance (TX IQ) and phase noise due to the transmit quadrature mixers and the nonlinear behavior of the transmit PA. On the receive

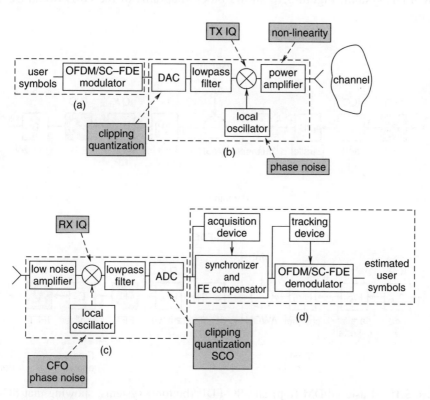

Figure 5.17 Non-ideality in direct conversion transmitters and receivers: (a) transmit digital transceiver; (b) transmit analog front-end; (c) receive analog front-end; (d) receive digital transceiver [15]. Reproduced by permission of © 2008 John Wiley & Sons, Ltd.

side, we recognize the receive IQ imbalance (RX IQ), CFO and phase noise due to the receive quadrature mixers, the clipping, quantization and sample clock offset (SCO) of the ADC. Without loss of generality, we assume that the CFO and SCO are due to the receiver, whereby the transmitter clock and LO are taken as references.

5.5.2 SC with Frequency-Domain Equalization

5.5.2.1 System Model

SC-FDE can be seen as a special case of linearly pre-coded OFDM [16, 17]: if, in an OFDM system, an FFT matrix is used to linearly pre-code the OFDM transmitted symbol blocks, it cancels the effect of the IFFT at the transmitter. An IFFT must then be inserted at the output of the equalizer in the receiver to compensate for the linear pre-coding with an FFT matrix at the transmitter. The resulting system is an SC-FDE system. Figure 5.18 shows block diagrams of the OFDM and SC-FDE systems.

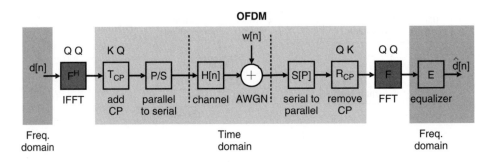

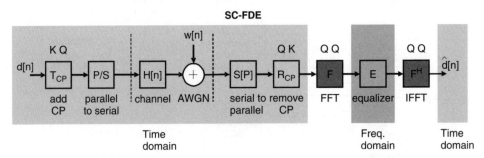

Figure 5.18 Basic OFDM (top) and SC-FDE (bottom) systems, showing that SC-FDE is equivalent to an OFDM system with a linear FFT pre-coding [13]. Reproduced by permission of © 2008 IEEE.
Note: AWGN = additive white Gaussian noise.

SC-FDE is an interesting alternative approach to OFDM. Both benefit from the same low-complexity multi-path channel equalization in the frequency domain. However, they feature different properties when they are implemented in actual wireless systems. SC-FDE benefits from a lower PAPR than OFDM because no signal pre-coding with an IFFT is performed at the transmitter. Another difference is the computational complexity required at the transmitter and receiver. In the case of OFDM, one FFT or IFFT operator is performed on both sides of the link. In the case of SC-FDE, no FFT/IFFT operator is performed on the transmit side while two FFT/IFFT operators are performed on the receive side of the link.

The simplified block input–output of SC-FDE in discrete time can be expressed as follows [15]:

$$\hat{\mathbf{d}}[n] = \mathbf{F}^H \mathbf{E} \mathbf{F} \mathbf{R}_{\text{CP}}(\mathbf{H}_t \mathbf{T}_{\text{CP}} \mathbf{d}[n] + \mathbf{w}[n])$$

$$= \mathbf{F}^H \mathbf{E} \mathbf{H}_f \mathbf{F} \mathbf{d}[n] + \mathbf{F}^H \mathbf{E} \mathbf{F} \mathbf{R}_{\text{CP}} \mathbf{w}[n] \tag{5.13}$$

where $\mathbf{d}[n]$ is the Q-vector of time-domain symbols in the nth block, $\mathbf{F}$ and $\mathbf{F}^H$ are the direct and inverse $Q \times Q$ Fourier transform matrices, $\mathbf{T}_{\text{CP}}$ and $\mathbf{R}_{\text{CP}}$ are matrices that, respectively, add and remove the cyclic prefix, $\mathbf{E}$ is the frequency-domain equalizer matrix, which is diagonal, $\mathbf{H}_t$ is the time-domain channel convolution matrix, which is a Toeplitz matrix, and $\mathbf{H}_f$ is the diagonal frequency-domain channel matrix equal to $\mathbf{F} \mathbf{R}_{\text{CP}} \mathbf{H}_t \mathbf{T}_{\text{CP}} \mathbf{F}^H$. The second equality in Equation (5.13) clearly shows that the SC-FDE system is equivalent to an OFDM system with a linear FFT pre-coding. It should be noted that the frequency-domain channel matrix $\mathbf{H}_f$ is diagonal because of the insertion and removal of the cyclic prefix. Indeed, the product $\mathbf{R}_{\text{CP}} \mathbf{H}_t \mathbf{T}_{\text{CP}}$ results in a circulant matrix, hence it is diagonalized by Fourier matrices in the product $\mathbf{F} \mathbf{R}_{\text{CP}} \mathbf{H}_t \mathbf{T}_{\text{CP}} \mathbf{F}^H$. It is this property that enables the channel in the frequency domain to be equalized by means of a diagonal matrix $\mathbf{E}$ (one coefficient per sub-carrier).

The $Q \times Q$ equalizer matrix $\mathbf{E}$ is diagonal with its diagonal elements equal to

$$e_{qq}^{\text{MMSE}} = \frac{\sigma_d^2 h_{f,q}^*}{\sigma_w^2 + \sigma_d^2 |h_{f,q}|^2} \tag{5.14}$$

for the minimum mean-square error (MMSE) equalizer, and to

$$e_{qq}^{\text{ZF}} = \frac{h_{f,q}^*}{|h_{f,q}|^2} \tag{5.15}$$

for the zero-forcing (ZF) equalizer ($h_{f,q}$ is the frequency-domain channel coefficient at sub-carrier q, σ_d^2 is the average symbol power and σ_w^2 is the average noise variance). These equations are valid for both OFDM and SC-FDE. It should be noted that the MMSE equalizer in Equation (5.14) tends to the ZF at high SNR since σ_w^2 then tends to zero.

5.5.2.2 Cyclic Prefix Versus Training Sequence for SC-FDE

Interestingly, it is possible to make a cyclic extension of the block transmission by inserting a known training sequence (TS) instead of a cyclic prefix. Both methods are illustrated in Figure 5.19. There are some differences between the two methods:

- In the case of the cyclic prefix, the CP must be discarded on the receiver side; an FFT of size N is computed on the remainder of the received block. In the case of the training sequence, however, the FFT, equalization and IFFT must be computed on the received block (thus including the TS) and the TS must be discarded afterwards in the time domain.
- There is a slight throughput advantage for CP because the throughput is proportional to $N/(N + N_{CP})$ for CP-SC-FDE while it is proportional to $(N - N_{TS})/N$ for TS-SC-FDE, which is lower for $N_{CP} = N_{TS}$. However, this advantage is offset by the fact that the TS can be used as pilots whereas, in CP-SC-FDE, pilots must be inserted in the data part of the block for tracking.
- If pilot symbols are inserted in CP-SC-FDE, it is most advantageous to insert them in the part of the block that is copied in the CP. In that case, each pilot symbol is repeated "for free" in the CP.
- The power consumption of the FFT/IFFT is higher for TS than for CP because, for TS, the FFT/IFFT must run 100% of the time while for CP it must run $100N/(N + N_{CP})$ percent of the time.

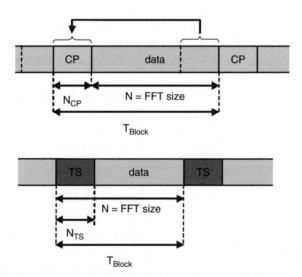

Figure 5.19 Two methods of cyclic extension for block communication: cyclic prefix (top) and training sequence (bottom).

5.6 Digital Baseband Processing

Based on the description of the IEEE 802.15.3c SC PHY layer in Section 5.4.2, a block diagram of the signal processing in the transmitter is shown in Figure 5.20. This block diagram assumes that the digital symbol stream is up-sampled by a factor of 2 and digitally pulse shaped before the DAC. An alternative implementation is to feed the DAC at the symbol rate and to do the pulse shaping with the analog filter following the DAC. This requires a tighter design of the analog filter but results in a more economical implementation in terms of hardware and power consumption.

A possible signal processing flow to receive the IEEE802.15.3c PHY bursts transmitted over a multi-path channel is shown in Figure 5.21. Note that this processing flow can be simplified in (nearly) LOS channels by opting for time-domain equalization. In this case, the channel estimation is simpler (only one complex coefficient must be estimated) and the FFT and IFFT may be bypassed.

5.6.1 Burst Detection and Rough Timing/CFO Acquisition

The first operation needed to process a burst is to detect the burst and acquire the timing and CFO information. In fact, the detection process must run continuously, and must therefore be implemented with a simple algorithm and at low power. The

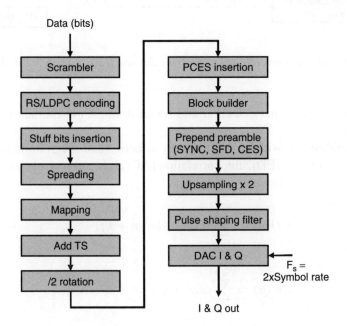

Figure 5.20 Transmitter signal processing for the IEEE 802.15.3c SC PHY.

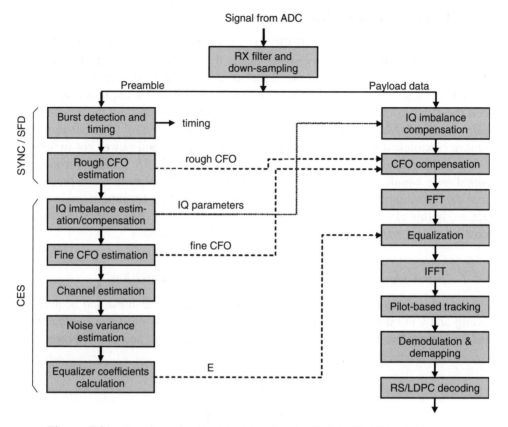

Figure 5.21   Receiver signal processing for the IEEE 802.15.3c SC PHY.

block diagram of the burst detection and rough timing/CFO estimator is shown in Figure 5.22.

Three algorithms are widely used for burst detection and rough timing/CFO acquisition: energy detection (ED), autocorrelation (AC) and cross-correlation (XC). They are described by the following equations:

$$ED(n) = \sum_{k=0}^{N_d-1} r(n+k)r^*(n+k), \tag{5.16}$$

$$AC(n) = \sum_{k=0}^{N_d-1} r(n+k)r^*(n+k-N_S), \tag{5.17}$$

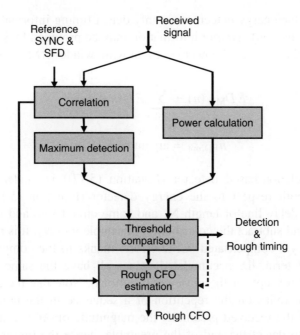

Figure 5.22 Burst detection and rough timing/CFO estimation.

$$XC(n) = \sum_{k=0}^{N_d-1} r(n+k)p^*(k), \tag{5.18}$$

where $r(n)$ is the received preamble sequence corrupted by the multi-path channel and additive white Gaussian noise, N_d is the number of samples over which the detection is calculated and N_S is the period of the SYNC sequence. In practice, the detector uses either $ED(n)$, $AC(n)$ or $XC(n)$ (or a function thereof) for the detection and a binary hypothesis test is performed. The probability density function (PDF) of the detector output when noise only is present is needed and the false alarm rate must be specified, so that a threshold can be computed for the detector [18]. The detection probability can then be computed by means of the PDF of the detector output when noise and signal are present [15].

The properties of these detection mechanisms vary widely:

- The energy detector is the simplest to implement and is insensitive to CFO. However, since it does not exploit any knowledge of the received preamble, it will react to any energy entering the baseband, including interference, and create false alarms. For this reason, it is not often used alone for burst detection.

Furthermore, the energy detector can only detect timing information (in addition to the burst detection). Timing information may be provided by searching for the maximum of the integrated power over a moving window (MW) of length N_d as:

$$ED_{MW}(n) = \sum_{m=0}^{N_d} ED(n+m),$$

$$\hat{n}_{ED,opt} = \arg\max_n(|ED_{MW}(n)|).$$

- The autocorrelation-based detector (Equation (5.17)) is moderately simple to implement. With respect to the energy detector (Equation (5.16)), it requires an additional delay line of length N_S and some circuitry to add the contribution of sample n and subtract the contribution of sample $n - N_d$. It is relatively insensitive to (not yet compensated) CFO errors: thanks to the complex conjugation in the second term, the accumulated values all have the same phase and add up coherently (except for the noise contributions). Note that the autocorrelation-based detector relies on the repetition of a sequence in the preamble to work. At the end of the received preamble, the magnitude of $AC(n)$ stops increasing. This is used to detect the end of the preamble, hence the timing information is known. Several detectors have been proposed to locate this peak, but we will use a max function since Fort et al. [19] suggest it is a good approximation of the maximum likelihood solution. Thus, the timing offset is estimated as follows:

$$\hat{n}_{AC,opt} = \arg\max_n(|AC(n)|).$$

Interestingly, the AC provides a simple means to estimate of the CFO as follows:

$$\hat{\Delta}_f = \frac{1}{2\pi N_S T}\angle[AC(\hat{n}_{AC,opt})],$$

where T is the inverse of the symbol rate. This estimator measures the phase rotation of the received signal over a duration $N_S T$. Note that the unambiguous estimation range of this estimator is equal to the inverse of the period of the sequence. Specifically, the unambiguous range is in the interval

$$\left[\frac{-1}{2N_S T}; \frac{+1}{2N_S T}\right] = [-6.75\,\text{MHz}; +6.75\,\text{MHz}].$$

Practical systems often have to compensate for a 40 ppm CFO, to account for 20 ppm crystal inaccuracy on both the TX and RX sides. At 60 GHz, this 40 ppm CFO amounts to 2.4 MHz, which is well within the acquisition range.

- The cross-correlation-based detector (Equation (5.18)) is more complex to implement because it requires N_d complex multiply – sum operations for each incoming sample. The optimum timing is found by looking for the highest peak in the correlator output:

$$\hat{n}_{\text{XC,opt}} = \arg\max_n(|XC(n)|).$$

Note that, as opposed to the AC output, the XC output is affected by the multi-path channel. In the ideal case, assuming zero-correlation side-lobes, zero carrier frequency offset and no noise, the cross-correlator output is equal to the channel impulse response (in practice, none of these assumptions is correct and only a distorted channel impulse response is available). Hence, multiple peaks may be present at the cross-correlator output and some ad-hoc method is needed to cope with multi-path channels. See [15, 19] for a detailed treatment of this. If a repetition is present in the preamble, a good strategy is to search for the periodic peaks in the magnitude of $XC(n)$ and find the time index that maximizes the sum of these peaks. Just as for the autocorrelation, the phase rotation between the cross-correlation peaks can be used to derive the CFO $\hat{\Delta}_f$.

For both the AC and the XC methods, it is common to normalize the correlator output by the value of the signal energy to ease the thresholding operation. The actual threshold must be calculated based on the statistics of the signal under the "noise only" condition and the "signal + noise" condition. It is important to note that the auto- and cross-correlation methods each have their own advantages in terms of performance: the autocorrelator uses a shifted replica of itself as a reference, hence it is rather insensitive to channel effects but is more sensitive to noise because noise is also present in the reference. The cross-correlator uses the known transmitted sequence as a reference, hence is more sensitive to channel distortion but has a reference waveform free of noise. Figure 5.23 shows the detection performance of the auto- and cross-correlation in CM15 channels [20]. The figure shows the percentage of successful burst detections for different values of E_b/N_0 applied on the SYNC sequence. It can be seen that the cross-correlation performs better for low values of E_b/N_0. With autocorrelation, noise terms are multiplied. Although the multiplied noise terms are uncorrelated, the variance of their product is equal to σ_n^4, where σ_n^2 is the noise variance of the individual noise terms. For small values of E_b/N_0 the term in σ_n^4 starts to dominate, rendering an effective detection impossible with autocorrelation.

Performance of the CFO estimation based on the cross-correlation is shown in Figure 5.24. The figure shows the normalized MSE of the estimated relative CFO (i.e. the MSE divided by the square of the true relative CFO). The simulation was done with 1000 realizations and a relative CFO of 40 ppm.

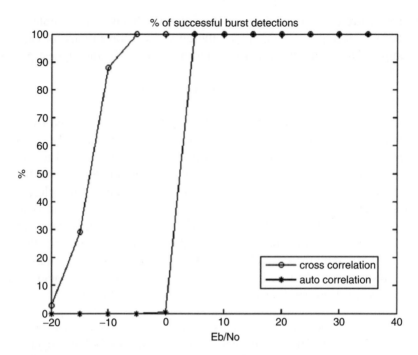

Figure 5.23   Burst detection and rough timing/CFO estimation.

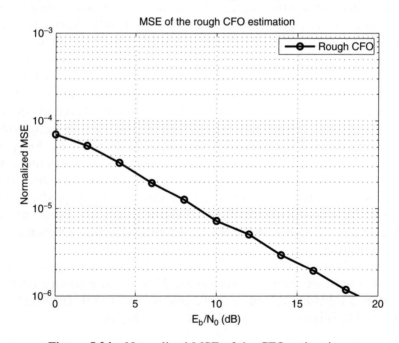

Figure 5.24   Normalized MSE of the CFO estimation.

5.6.2   Joint Fine CFO and Channel Estimation Without I/Q Imbalance

Having acquired timing and rough CFO estimation, the next step is to estimate the channel. If no I/Q imbalance is present in the receiver, the residual CFO must be estimated together with the channel response based on the received CES sequence. The CES field, used for channel estimation, consists of the sequence $[\mathbf{b}_{128} \ \mathbf{b}_{256}$ $\mathbf{a}_{256} \ \mathbf{b}_{256} \ \mathbf{a}_{256}]$, where the leftmost sequence, $\mathbf{a}_{256}$, is first in time and where the first transmitted bit is the rightmost bit. For the sake of clarity and to follow a more conventional notation, in this section we will express this as follows: the CES consists of the sequence $[\mathbf{a}_{256} \ \mathbf{b}_{256} \ \mathbf{a}_{256} \ \mathbf{b}_{256} \ \mathbf{b}_{128}]$ where the first transmitted bit is the leftmost bit of $\mathbf{a}_{256}$. Since $\mathbf{a}_{256} = [\mathbf{b}_{128} \ \mathbf{a}_{128}]$ and $\mathbf{b}_{256} = [-\mathbf{b}_{128} \ \mathbf{a}_{128}]$, we have the following properties:

$$CES = [\mathbf{a}_{256} \ \mathbf{b}_{256} \ \mathbf{a}_{256} \ \mathbf{b}_{256} \ \mathbf{b}_{128}]$$

$$= [\mathbf{b}_{128} \ \mathbf{a}_{128} \ -\mathbf{b}_{128} \ \mathbf{a}_{128} \ \mathbf{b}_{128} \ \mathbf{a}_{128} \ -\mathbf{b}_{128} \ \mathbf{a}_{128} \ \mathbf{b}_{128}]$$

$$= [\mathbf{b}_{128} \ \mathbf{a}_{512} \ \mathbf{a}_{512}]$$

Hence, the CES sequence consists of the sequence $\mathbf{a}_{512} = [\mathbf{a}_{256} \ \mathbf{b}_{256}] = [\mathbf{a}_{128} \ -\mathbf{b}_{128}$ $\mathbf{a}_{128} \ \mathbf{b}_{128}]$ repeated twice and preceded by the cyclic prefix $\mathbf{b}_{128}$. This is a cyclic extension of only 128 samples, which is supposed to be long enough for most 60 GHz channels. For notational purposes, we will denote the components of the received CES sequence by $\mathbf{p}$ and $\mathbf{q}$, where $\mathbf{p}$ stands for a received $\mathbf{a}$ sequence and $\mathbf{q}$ for a received $\mathbf{b}$ sequence. Note that the received $\mathbf{p}$ and $\mathbf{q}$ sequences are equal to the transmitted $\mathbf{a}$ and $\mathbf{b}$ sequences after convolution with the channel and addition of the receiver noise:

$$\mathbf{p}(n) = \mathbf{a}(n) \otimes \mathbf{h}(n) + \mathbf{n_a}(n)$$

$$\mathbf{q}(n) = \mathbf{b}(n) \otimes \mathbf{h}(n) + \mathbf{n_b}(n)$$

The correlation of the transmitted CES with $\mathbf{a}_{512}^{*}$ shows two length-128 zero-correlation zones after the main peak (marked in Figure 5.25). Hence, after convolution with the channel, the channel impulse response can be directly extracted from both zero-correlation zones and averaged to reduce the impact of the additive noise by 3 dB. The impact of residual CFO and possible I/Q imbalance must be compensated before estimating the channel.

A good strategy for carrying out this joint channel and CFO estimation is the following process on the received CES sequence:

- Discard the cyclic prefix $\mathbf{q}_{128}$.
- Compute the autocorrelation of the first part of the received CES with the second part and estimate the residual CFO as follows:

$$\hat{\Delta}_f = \frac{1}{2\pi 512 T} \angle [\mathbf{p}_{512} \otimes \mathbf{p}_{512}^{*}].$$

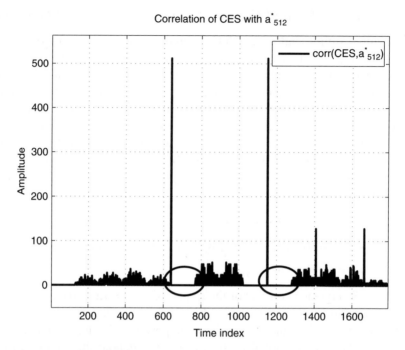

Figure 5.25 Matched filtering of CES with $\mathbf{a}_{512}$. Two length-128 zero-correlation zones (indicated by ellipses) can be used for direct estimation of the channel impulse response.

- Correct the second part of the received CES by means of the estimated CFO, i.e. multiply by $\exp(-j2\pi)\hat{\Delta}_f kT$, where k spans the range from 1 to the length of the sequence to be compensated.
- For time-domain channel estimation, extract the impulse response by element-wise averaging of the impulse responses measured in the two zero-correlation zones.
- For frequency-domain channel estimation, compute the length-512 DFT of the two element-wise averaged $\mathbf{p}_{512}$ sequences. A length-512 DFT is needed because the block length used for FDE is 512. In the frequency domain, divide element-wise the resulting vector by the DFT of the $\mathbf{a}_{512}$ sequence.

5.6.3 Joint Estimation of Fine CFO, Channel and I/Q Imbalance

5.6.3.1 Impact of I/Q Imbalance

When I/Q imbalance is present, the complex received signal at baseband, including the received preamble, is distorted. If the receiver I and Q branches do not have equal gain or are not exactly in quadrature, signal distortion results. These

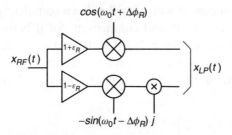

Figure 5.26 Receiver I/Q imbalance equivalent diagram (amplitude offset: ε_R; phase offset: $\Delta\phi_R$) [15]. Reproduced by permission of © 2008 John Wiley & Sons, Ltd.

differences can be represented by different gains ($1 + \varepsilon_R$ and $1 - \varepsilon_R$) and different phases ($+\Delta\phi_R$ and $-\Delta\phi_R$), as illustrated in Figure 5.26. The distorted representation of the signal is then, using ε_R and $\Delta\phi_R$ for the gain and phase imbalances and assuming no CFO and no noise:

$$y(t) = \alpha x(t) + \beta x^*(t), \tag{5.19}$$

in which

$$\alpha = \cos(\Delta\phi_R) - j\varepsilon_R \sin(\Delta\phi_R), \tag{5.20}$$

$$\beta = \varepsilon_R \cos(\Delta\phi_R) + j \sin(\Delta\phi_R). \tag{5.21}$$

Equation (5.19) reveals that the resulting low-pass signal $y(t)$ consists of the ideal signal $x(t)$ plus a scaled version of its complex conjugate $x^*(t)$. Because of the complex conjugation, the positive part of the spectrum of $X(f)$ interferes with the negative part of its spectrum and vice versa. Note that for reasonable values of ε_R and $\Delta\phi_R$, the magnitude of α is close to one and the magnitude of β is close to zero.

5.6.3.2 Compensation of I/Q Imbalance

It is easy to show that the knowledge of β/α^* is sufficient to compensate for the I/Q imbalance. Indeed, we can compensate the received signal $y(t)$ as follows:

$$y_{\text{comp}}(t) = y(t) - \frac{\beta}{\alpha^*}y^*(t)$$

$$= \alpha x(t) + \beta x^*(t) - \frac{\beta}{\alpha^*}(\alpha^*x^*(t) + \beta^*x(t))$$

$$= \frac{|\alpha|^2 - |\beta|^2}{\alpha^*}x(t) \tag{5.22}$$

We observe that the component with $x^*(t)$ has been completely removed in Equation (5.22). So, if we estimate β/α^* and compensate for it before the channel estimation, the image interference due to I/Q imbalance is completely removed and the remaining scaling will be compensated for by the channel estimation/equalization. Note that the magnitude of this scaling is close to one, so it has insignificant impact on the SNR.

5.6.3.3 Estimation of the I/Q Imbalance Parameters

The periodic structure of the preamble can be used to make the estimate of the CFO and I/Q parameters independent of the channel knowledge. The method introduced here is described in detail in [21]. Although the mathematical treatment of this method is quite involved, its implementation is fairly straightforward. We will highlight the main idea of the method here and refer the reader to [21] for the details. The parameters to estimate are the phase rotation ϕ between the first and second part of the received preamble and the ratio $\bar{\beta} = \beta/\alpha^*$.

First, exploiting the repetition in the preamble, the second part $\mathbf{y}_2$ of the received preamble is expressed as a function of the first received part $\mathbf{y}_1$, affected by CFO and I/Q imbalance:

$$\mathbf{y}_2 = \lambda(\phi, \bar{\beta})\, \mathbf{y}_1 + \mu(\phi, \bar{\beta})\, \mathbf{y}_1^* + \mathbf{n}, \tag{5.23}$$

in which the parameters $\lambda(\phi, \bar{\beta})$ and $\mu(\phi, \bar{\beta})$ are given by

$$\lambda(\phi, \bar{\beta}) := \cos\phi - j\,\sin\phi\,\frac{1 + |\bar{\beta}|^2}{1 - |\bar{\beta}|^2}, \tag{5.24}$$

$$\mu(\phi, \bar{\beta}) := 2j\,\sin\phi\,\frac{\bar{\beta}}{1 - |\bar{\beta}|^2}. \tag{5.25}$$

Second, the maximum likelihood (ML) function for the parameters to estimate (CFO, I/Q) is derived:

$$\Lambda(\phi, \bar{\beta}) := \log p\left(\mathbf{y}_2|\mathbf{y}_1, \phi, \bar{\beta}\right) \tag{5.26}$$

$$= -\frac{Q}{2}\log\left(2\pi\sigma_n^2(\phi, \bar{\beta})\right)$$

$$- \frac{1}{2\sigma_n^2(\phi, \bar{\beta})}\left(\mathbf{y}_2 - \lambda(\phi, \bar{\beta})\mathbf{y}_1 - \mu(\phi, \bar{\beta})\mathbf{y}_1^*\right)^H$$

$$\cdot \left(\mathbf{y}_2 - \lambda(\phi, \bar{\beta})\mathbf{y}_1 - \mu(\phi, \bar{\beta})\mathbf{y}_1^*\right), \tag{5.27}$$

where the noise vector **n** is given by

$$\mathbf{n} := -\lambda(\phi, \bar{\beta})\, \mathbf{z}_0 - \mu(\phi, \bar{\beta})\, \mathbf{z}_0^* + \mathbf{z}_1. \tag{5.28}$$

The variance of the noise vector can be shown to be equal to

$$\sigma_n^2(\phi, \bar{\beta}) = \sigma_{n_r}^2(\phi, \bar{\beta}) + \sigma_{n_i}^2(\phi, \bar{\beta})$$

$$= \sigma_z^2 \left(2 - |\bar{\beta}|^2 + 8\,(\sin\phi)^2 \, \frac{|\bar{\beta}|^2}{1 - |\bar{\beta}|^2} \right), \tag{5.29}$$

which shows that the variance of the noise samples depends on the CFO and I/Q imbalance.

Third, because solving for the ML estimate by means of this function is not tractable, a simplification is needed. Because the I/Q mismatch parameter $\bar{\beta}$ is small for actual analog front-ends ($\bar{\beta} \ll 1$), we can approximate the likelihood function by making a second-order approximation. The Equations (5.24), (5.25) and (5.29) are approximately equal to

$$\lambda(\phi, \bar{\beta}) \approx e^{-j\phi} - 2\,j\,\sin\phi\,|\bar{\beta}|^2, \tag{5.30}$$

$$\mu(\phi, \bar{\beta}) \approx 2\,j\,\sin\phi\,\bar{\beta}, \tag{5.31}$$

$$\sigma_n^2(\phi, \bar{\beta}) \approx 2\sigma_z^2 \, \frac{1}{1 + (\frac{1}{2} - 4\,(\sin\phi)^2)\,|\bar{\beta}|^2}. \tag{5.32}$$

These expressions are inserted into the likelihood function ($\Lambda(\phi, \bar{\beta})$), which is further simplified by neglecting the terms of order higher than 2.

Fourth, in order to solve for the two parameters to be estimated (ϕ and β), the expectation–maximization (EM) algorithm is used to estimate iteratively the CFO and I/Q imbalance one after each other. The iterative EM algorithm is known to converge to the joint ML estimate of the parameters [22]. The EM algorithm is typically used to estimate a set of parameters based on a given observation when unknown random processes perturb the observation. In our case, the phase ϕ due to the CFO is estimated based on the two received vectors ($\vec{y}_1$ and $\vec{y}_2$). The I/Q imbalance $\bar{\beta}$ alters the estimation of ϕ. The algorithm is initialized by a first estimate of the CFO (ϕ) identical to the autocorrelation method described earlier. Then the algorithm iterates between individual estimations of the I/Q parameters ($\bar{\beta}$) and CFO (ϕ) until it converges.

When the I/Q imbalance parameters are very large ($\varepsilon_R > 5\%$ and $\Delta\phi_R > 5°$), the estimation starts to degrade. A good solution in this case consists of doing a first EM estimation and then compensating the received CES for the estimated CFO and I/Q. Then, simple independent ML estimates of the CFO and I/Q based

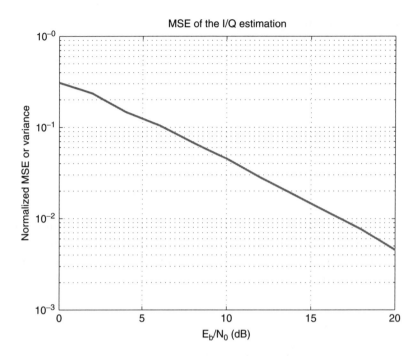

Figure 5.27   Normalized MSE of the I/Q parameter estimation.

on the received and compensated CES are sufficient. Figure 5.27 shows the performance of the described EM-based I/Q estimation algorithm, averaged over 500 channel realizations. The I/Q parameters were $\varepsilon_R = 5\%$ and $\Delta\phi_R = 5°$. Once the frequency-domain coefficients $H(k)$ and the noise variance are known, the equalizer coefficients can be readily calculated. For the ZF and the MMSE equalizer, the equalizer coefficients are given by Equations (5.14) and (5.15), respectively.

5.6.3.4   Tracking of CFO and Phase Noise

Because the CFO estimation is not perfect, especially at low SNR, and because of the presence of phase noise, it is necessary to track the residual CFO and the phase noise during the reception of a burst. The TS used to cyclically extend the transmitted blocks can be used for this purpose (Figure 5.28). The operation is as follows:

- Each block (of length 512 with the TS) is converted to the frequency domain and equalized.
- The equalized frequency-domain blocks are converted back to the time domain.

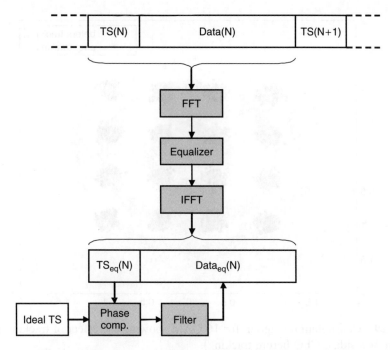

Figure 5.28 Tracking based on the TS after equalization.

- The TS, which occupies the first 64 samples of the blocks, is compared against the ideal TS to derive the phase error.
- The phase error compensation is applied to the data part of the burst ($512 - 64 =$ 448 samples).

Figure 5.29 illustrates the behavior of the tracking for 16-QAM at 20 dB SNR, when the residual CFO is 2 ppm before the tracking loop. The value of residual CFO for the simulated SNR is rather high and should be much smaller after coarse and fine CFO estimation. It can be observed that the constellation points after tracking are perfectly centered on the locations of an ideal 16-QAM constellation.

5.6.4 Time-Domain Equalization, Despreading and Tracking

In LOS and near-LOS channels, it may not be necessary to perform frequency-domain equalization. Indeed, in such channels, the main tap of the channel impulse response (obtained by means of the procedure with the matched filter outlined in Section 5.6.2) can be used to perform a single-tap time-domain equalization. To decide between time- and frequency-domain equalization, a criterion based on the

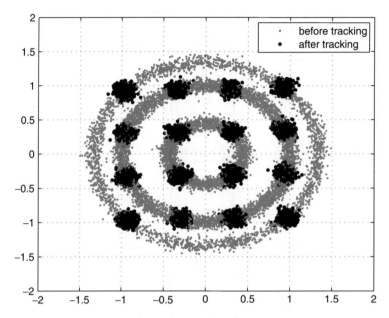

Figure 5.29 Constellation diagram for 16-QAM showing the effect of tracking (assuming 2 ppm of residual CFO before tracking).

ratio of the energy of the strongest path to the sum of the energies of the other taps,

$$r_{\text{tap}} = \frac{|h_{n_{\max}}|^2}{\sum_{\substack{n=0 \\ n \neq n_{\max}}}^{N-1} |h(n)|^2},$$ (5.33)

can be used ($n_{\max}$ is the index of the strongest tap).

In time-domain equalization, the processing must be done in a different order to accommodate the possible despreading. Since the spread chips have been affected by the $\pi/2$ rotation, the $\pi/2$ derotation must be applied before the despreading. It is more economical to perform the time-domain equalizer on the despread chips. Finally, if no TS was added to the blocks, a decision-directed CFO/phase noise tracking loop must be used, whereby the outputs of the hard decision are used as reference to the phase comparator. This whole process is illustrated in Figure 5.30.

In order to assess the impact of PA nonlinearity, we use a modified Rapp model taking into account AM-AM and AM-PM conversion as discussed in Chapter 4, since this is more realistic for microwave and millimeter-wave PAs. The backoff is expressed relative to the 1 dB compression point, which is the point where the gain drops by 1 dB with respect to the gain at low power. Coded and uncoded BER performance for QPSK and 16-QAM are shown in Figures 5.31 and 5.32, respectively. In both cases, the coded BER were simulated in nearly LOS channels

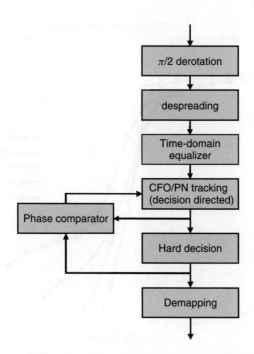

Figure 5.30   Tracking combined with despreading.

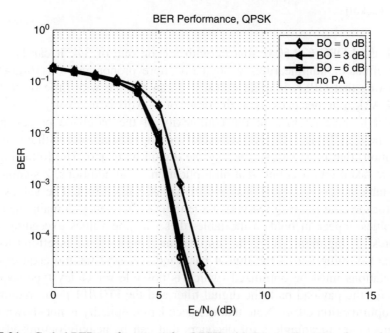

Figure 5.31   Coded PER performance for QPSK with PA nonlinearity, for various levels of backoff.

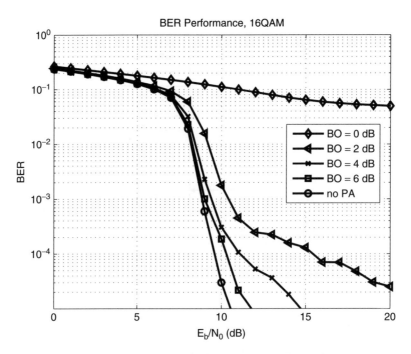

Figure 5.32 Coded PER performance for 16-QAM with PA nonlinearity, for various levels of backoff.

and with the LDPC(672,504), that is, with a code rate of 3/4. It can be observed that the effect of nonlinearities is logically more pronounced on 16-QAM. It is worth noting that QPSK is quite robust and can operate with little or no backoff. For 16-QAM, a higher backoff of the order of 2–3 dB is necessary.

In order to give an idea of the complexity of the baseband processing for a SC, we show in Table 5.10 the number of raw operations (i.e. without significant optimization) that must be executed by the digital baseband. The following operations were taken into account: real multiplications, real additions, 2's complement, look-up-table access, comparisons, real divisions, square root. The values are calculated for the case of 16-QAM with TS length 64, frequency-domain equalization. Some values appear in bold in the table: these are the values exceeding 10 giga-operation/s, to highlight where the implementer must focus its efforts to reduce the power consumption of the digital baseband. Clearly, for the acquisition part, the correlations must be optimized (which is possible for Golay type sequences), whereas for the payload part, the digital filter and the FFT/IFFT will require a significant optimization effort. Note that the decoder complexity is not shown because raw numbers are meaningless for decoders that can be heavily optimized (e.g. by means of approximations, parallelization and stop criteria in the iterations).

Table 5.10 Complexity of the signal processing functions of the SC PHY. Reproduced by permission of © 2008 IEEE

RECEIVER	Input unit	Input rate G/s	Output unit	Output rate G/s	Real Mult in G/s	Real Add in G/s	2s Comp. in G/s	LUT real out in G/s	Compar. in G/s	Real Div. in G/s	Sqrt in G/s
Digital filter + downsampling	Complex	3.456	Complex	1.7280	**86.4000**	**82.9440**					
Synchronization											
Correlation	Complex	1.7280	Complex	1.7280	3.4560	**883.0080**	1.7280				
Power calculation					3.4560	1.7280					
Maximum detection									1.7280		
Threshold comparison					0.2025	1.9980	0.0473	0.0068	0.0068	0.0068	
Rough CFO estimation					0.0270	0.0068		0.0068	0.0068	0.0068	0.0068
Estimation											
Pos & Neg CFO compensation	Complex	1.7280	Complex	1.7280	15.5520	13.8240	1.7280	3.4560			
Correlations	Complex	1.7280	Complex	0.6480		**883.0080**	2.1600				
LSE estimator					5.2329	7.8165	0.6522			0.0017	
P & Q calculation					0.8640	1.2960	0.2160				
Fine CFO estimation					0.4354	1.2952	0.1080	0.0008		0.0008	0.0008
Channel estimation					0.4320	0.6480					
Noise Power estimation					0.4337	0.6472					
Channel DFT					15.5520	15.5520					
MMSE equalizer					3.8880	4.7520	0.8640			0.4320	
Data processing											
Rx IQ compensation	Complex	1.7280	Complex	1.7280	6.9120	10.3680	1.7280				
CFO compensation	Complex	1.7280	Complex	1.7280	8.6400	6.9120		3.4560			
DFT	Complex	1.7280	Complex	1.7280	**62.2080**	**62.2080**					
Equalization	Complex	1.7280	Complex	1.7280	6.9120	6.9120					
IDFT	Complex	1.7280	Complex	1.7280	**62.2080**	**62.2080**					
Tracking + TS removal	Complex	1.7280	Complex	1.5120							
Rotation calculation					0.8708	1.2926	0.0034			0.00675	0.003375
Rotation compensation					6.0480	6.0480					
Tx IQ compensation	Complex	1.5120	Complex	1.5120	12.0960	15.1200	1.5120				
pi/2 derotation	Complex	1.5120	Complex	1.5120			1.5120				
Demapping (hard)	Complex	1.5120	Bit	6.0480					9.072		
Demapping (soft)	Complex	1.5120	real	6.0480					9.072		

References

[1] Van Nee, R. and Prasad, R. (1999) *OFDM for Wireless Multimedia Communications*. Boston: Artech House.

[2] Clark, M.V. (1998) Adaptive frequency-domain equalization and diversity combining for broadband wireless communications. *IEEE Journal on Selected Areas in Communications*, **16**(8).

[3] Falconer, D., Ariyavisitakul, S., Benyamin-Seeyar, A. and Eidson, B. (2002) Frequency domain equalization for single-carrier broadband wireless systems. *IEEE Communications Magazine*, **40**(4), 58–66.

[4] Benvenuto, N. and Tomasin, S. (2002) On the comparison between OFDM and single carrier with a DFE using a frequency domain feed-forward filter. *IEEE Transactions on Communications*, **50**(6), 947–955.

[5] Yong, S.K. and Singh, H. (2009) OFDM system design for 60 GHz high-definition video applications. *IMS Workshop on System-Level Design and Implementation of Gb/s 60 GHz Radios*.

[6] Goeckel, D.L. and Ananthaswamy, G. A comparison of single-carrier and multi-carrier methodologies for wireless communications from a coding, modulation, and equalization perspective. http://www-unix.ecs.umass.edu/goeckel/ofdm.html.

[7] Sobel, D. (2004) Opportunities and challenges in 60 GHz wideband wireless system design. Berkeley Wireless Research Center Summer Retreat Presentations.

[8] Rapp, C. (1991) Effects of HPA-nonlinearity on a 4-DPSK/OFDM-signal for a digital sound broadcasting system. In *Proceedings of 2nd European Conf. Satellite Communications*, Liége, pp. 179–184, October.

[9] Pollet, T., Van Bladel, M. and Moeneclaey, M. (1995) BER sensitivity of OFDM systems to carrier frequency offset and phase noise. *IEEE Transactions on Communications*, **43**(234).

[10] Chiueh, T.-D. and Tsai, P.-Y. (2007) *OFDM Baseband Receiver Design for Wireless Communications*. Chichester: John Wiley & Sons, Ltd.

[11] Lee, C. and Liu, S.-I. (2007) A 58-to-60.4 GHz frequency synthesizer in 90 nm CMOS. *IEEE International Solid-State Circuits Conference*, pp. 196–197, February.

[12] IEEE802.15.3c (2009). IEEE Standard for information technology – Telecommunications and information exchange between systems – Local and metropolitan area networks – Specific requirements. Part 15.3: Wireless medium access control (MAC) and physical layer (PHY) specifications for high rate wireless personal area networks (WPANs) Amendment 2: Millimeter-wave-based alternative physical layer extension. http://www.ieee802.org/15/

[13] Bourdoux, A., Nsenga, J., Thillo, W.V., Wambacq, P. and der Perre, L.V. (2008) Gbit/s radios @ 60 GHz: To OFDM or not to OFDM? *2008 IEEE 10th International Symposium on Spread Spectrum Techniques and Applications*, IEEE, pp. 560–565.

[14] Vaughan, R., Scott, N. and White, D.R. (1991) The theory of bandpass sampling. *IEEE Transactions on Signal Processing* **39**(9), 1973–1984.

[15] Horlin, F. and Bourdoux, A. (2007) *Digital Front-End Compensation for Emerging Wireless Systems*. John Wiley & Sons, Ltd.

[16] Czylwik, A. (1997) A comparison between adaptive OFDM and single carrier modulation with frequency domain equalization. In *1997 IEEE 47th Vehicular Technology Conference*, IEEE, vol. 2, pp. 865–869.

[17] Sari, H., Karam, G. and Jeanclaude, I. (1995) Transmission techniques for digital terrestrial TV broadcasting. *IEEE Communications Magazine*, **33**(2), 100–109.

[18] Kay, S. (1997) *Fundamentals of Statistical Signal Processing, Volume 2: Detection Theory*. Englewood Cliffs, NJ: Prentice Hall.

[19] Fort, A., Weijers, J., Derudder, V., Eberle, W. and Bourdoux, A. (2003) A performance and complexity comparison of auto-correlation and cross-correlation for OFDM burst synchronization. *2003 IEEE International Conference on Acoustics, Speech, and Signal Processing* (ICASSP '03), IEEE, pp. II341–II344.

[20] Yong, S.K. et al. (2007) TG3c Channel Modeling Sub-committee Final Report. IEEE 15-07-0584-01-003c, March.

[21] Horlin, F., Bourdoux, A., Lopez-Estraviz, E. and der Perre L.V. (2007) Low-complexity EM-based joint CFO and IQ imbalance acquisition. *Proceedings of IEEE International Conference on Communications* (ICC '07), pp. 2871–2876.

[22] Moon, T. (1996) The expectation maximization algorithm. *IEEE Signal Processing Magazine* **13**(6), 47–59.

6

60 GHz Radio Implementation in Silicon

Alberto Valdes-Garcia

6.1 Introduction

In recent years, the design of active and passive mm-wave components in general – and in the 60 GHz band in particular – has become a center of gravity for academic and industrial research. Within a period of six years, from the first 60 GHz building blocks integrated in silicon introduced in 2004 [1] to today, this field of research has quickly expanded, resulting in multiple examples of fully integrated radios and phased arrays.

Even though research and engineering activities involving mm-wave frequencies span almost a century, the recent availability of silicon processes that allow radio implementation at 60 GHz (90 nm CMOS and 0.13 µm SiGe BiCMOS) is arguably the single most important factor in fueling 60 GHz standardization and investment activities. The economic and technical advantages that only silicon integration can offer are expected to take 60 GHz technology to usage and business volumes comparable to those currently enjoyed by Bluetooth and Wi-Fi technologies.

It is therefore relevant to the design of 60 GHz systems to understand the capabilities and limitations that a silicon implementation of the physical layer (PHY) entails. Rather than discussing 60 GHz circuit design techniques in detail (which would require several chapters to cover material presented in depth elsewhere [2]), this chapter presents an overview of the current solutions, techniques and tradeoffs

60 GHz Technology for Gbps WLAN and WPAN: From Theory to Practice
Edited by Su-Khiong (SK) Yong, Pengfei Xia and Alberto Valdes Garcia
© 2011 John Wiley & Sons, Ltd

involved in the implementation of a high-data-rate 60 GHz radio in silicon from the radio frequency (RF) front-end to the mixed-signal (analog/digital) interface with a digital baseband chip. The objective of this chapter is to provide system architects, PHY designers, application development engineers, and other non-circuit-design experts with the appropriate knowledge and background on the crucial *non-digital* section of the PHY to make the right system-level assessments and tradeoffs when taking a 60 GHz system *from theory to practice*.

In the sections that follow, the discussion starts with an overview and analysis of the different silicon technologies available for the implementation of 60 GHz systems. Given that the link margin of a wireless system is strongly dependent on the receiver's noise figure (NF) and the transmitter's OP1dB, the performance of currently existing 60 GHz low-noise amplifier (LNA) and power amplifier (PA) solutions is next reviewed in detail. Radio architectures are then reviewed, considering frequency planning as well as implementation tradeoffs for single carrier (SC) and orthogonal frequency division multiplexing (OFDM) systems. The current state of the art in high-speed digital-to-analog converters (DACs), analog-to-digital converters (ADCs) and modulators as important system components is analyzed in the context of their application to Gbps systems. Finally, the challenges for the implementation of commercial mm-wave radios are outlined.

6.2 Overview of Semiconductor Technologies for 60 GHz Radios

In the early 1990s, the cutoff frequencies for silicon transistors (CMOS FETs and SiGe HBTs) were below 100 GHz and about an order of magnitude smaller than those achieved by the III-V semiconductor devices which dominated the RF and mm-wave regime. By 2005, however, both SiGe and CMOS transistors had cutoff frequencies exceeding 200 GHz. During this period the foundations for RF integrated circuit design were laid, multiple new design techniques and circuit topologies were constantly introduced, and the manufacturing of integrated passive components (RF inductors, capacitors and transmission lines) matured. All of these factors resulted in a quasi-exponential improvement of RF integrated circuit performance over time.

In turn, the steep upward slope of this performance trend has created high performance expectations for 60 GHz systems. During the course of 60 GHz technology development, predictions were made regarding potential improvements in circuit performance with technology scaling. For instance, some postulated that 45 nm technology could significantly enhance output power and power efficiency in 60 GHz PAs as compared to early results achieved in 90 nm technology. Without underestimating the innovation capabilities of the wireless engineering community, it is worth carrying out a "reality-check" on the silicon roadmap to know what to expect from future developments in silicon technology and the resulting implications for

60 GHz radio implementations. As will be clearly shown by the end of this chapter, much like projected improvements in modern multi-core microprocessors, a continuous increase in 60 GHz radio performance will be the result of careful system engineering rather than the automatic result of technology scaling.

For more than a decade, the International Roadmap for Semiconductors (ITRS) [3] has served as the main reference for tracking and projecting the evolution of the semiconductor industry. The biannual ITRS report and annual ITRS updates are written by experts from all major semiconductor companies as well as research institutions and universities. The ITRS outlook for the future considers both known production plans sustained by present research and development results and performance targets driven by the requirements of future applications. Before discussing the outlook for SiGe and CMOS, it is worth reviewing the definitions of the metrics employed to benchmark the RF performance of a given technology node:

Physical L_{gate}. This refers to the actual minimum gate length that a FET can have in a given CMOS process. Note that this parameter is different from the technology node name which is also given in nanometers or micrometers. In the early stages of RF CMOS, L_{gate} used to be half of the quoted technology dimension, but this is no longer the case for modern technologies. For example, L_{gate} in the 0.35 μm CMOS node was 0.2 μm, while in a typical 65 nm CMOS node, L_{gate} is 50 nm.

Supply voltage. For CMOS, this value is specified to ensure reliability of digital circuits. It must be considered carefully in the design of CMOS PAs, as will be discussed in Section 6.3.

BV_{ceo}. In a SiGe process, the maximum collector-emitter voltage is quoted for HBTs instead of a supply voltage.

Peak f_T. The frequency at which the current gain of a transistor (with an AC short circuit as a load) reaches unity is known as the current-gain cutoff frequency or f_T. This value is relatively independent of the specific device layout configuration but depends on the device bias and device size. The peak f_T is the highest f_T in a given technology node for an optimum device size and bias conditions.

Peak f_{MAX}. The frequency at which the power gain of a transistor (for power-matched source and load impedances) reaches unity is known as the power-gain cutoff frequency. It is strongly dependent on layout configuration.

NF_{min} at 60 GHz. The minimum 60 GHz noise figure that can be obtained for a device under optimum impedance matching and bias conditions. In practice, the LNA noise figure will be at least 1 dB higher than this value since (a) it is not always possible to implement the optimum source impedance match and (b) the passive components employed for impedance matching introduce loss which directly translates into an additional noise figure.

One of the key advantages of using CMOS technology for a 60 GHz radio is that it makes a single-chip implementation possible, which is desirable from at least

Table 6.1 ITRS performance outlook for RF CMOS FETs in the next five years

Year of production	2010	2011	2012	2013	2014
Physical L_{gate} [nm]	29	27	24	22	20
Supply voltage [V]	1	1	1	0.95	0.9
Peak f_T [GHz]	310*	330*	370**	400**	440**
Peak f_{MAX} [GHz]	380*	410*	460**	510**	560**
NF_{min} at 60 GHz [dB]	3.3	3.2	3.0	3.0	2.9

*Manufacturable solutions are known but not implemented.
**Manufacturable solutions are not known.

two perspectives. First, it would reduce the form factor of the complete 60 GHz solution. Second, the performance of the digital baseband would benefit from the use of the latest technology node. The implications of future technology scaling for the radio section are, however, not as straightforward. Table 6.1 presents the roadmap of CMOS development for the next five years in terms of RF performance. By 2014, the minimum gate-length for a CMOS FET is expected to scale down by about 30% and so the current gain and power cutoff frequencies are expected to improve by the same amount. This inverse proportional relationship between L_{gate} and f_T/f_{MAX} has been approximately constant over the last decade. Nevertheless, note that according to ITRS, manufacturable solutions to sustain this trend in transistor speed are not known for L_{gate} below 27 nm. The main reason for this concern is that at these deeply scaled dimensions and for the extremely high expected cutoff frequencies (entering the sub-millimeter wave regime starting at 300 GHz), second-order scaling effects and specific device layout properties become relevant. To overcome these limitations through layout and device engineering techniques is an active topic of research [4].

What are the implications of these scaling trends for circuits operating at 60 GHz? In general, the availability of devices with higher cutoff frequencies implies that compared to previous generations, a similar level of performance can be obtained at a relatively lower bias current. Nevertheless, given the strong dependency of cutoff frequencies on device geometry, and further that the nominal operating power supply voltage will remain essentially constant, the expected reduction in power consumption will be (without significant circuit design and/or device innovations) less than 20%.

It has been demonstrated experimentally that the current density at which the NF_{min} occurs for deep sub-micrometer CMOS remains essentially constant from node to node [5]. In addition, as improvements in NF_{min} in the next 5 years are expected to be marginal, 60 GHz LNA circuits are not expected to experience a significant power reduction with technology scaling. In the realm of PAs, an increasing f_{MAX} results in higher power added efficiency (PAE); however, a constant

Table 6.2 ITRS performance outlook for SiGe HBTs in the next five years

Year of production	2010	2011	2012	2013	2014
Emitter width [nm]	100	100	100	90	90
BV_{ceo} [V]	1.6	1.55	1.5	1.45	1.4
Peak f_T [GHz]	320	340	360*	380*	395*
Peak f_{MAX} [GHz]	350	370*	390*	410*	425**
NF_{min} at 60 GHz [dB]	1.9	1.7	1.5	1.4	1.3

*Manufacturable solutions are known but not implemented.
**Manufacturable solutions are not known.

(slightly decreasing, in fact) maximum supply voltage means that higher output power levels from 60 GHz CMOS transmitters will come from innovations in on-chip power combining techniques rather than technology scaling. Section 6.3 discusses the performance trends of LNAs and PAs in more detail.

Table 6.2 presents a similar roadmap for SiGe HBTs. In general, SiGe bipolar transistors are expected to maintain an advantage of about 1.5 dB in NF_{min} and 0.5 V in allowable voltage swing with respect to CMOS FETs over the next five years with a comparable increase in cutoff frequencies.

One of the special characteristics of III-V semiconductors (which dominated mm-wave applications until recently) is that their complex structure and low-volume business model allow for the development of different device variants in a given technology node. For example, a device can be engineered to achieve excellent noise performance (suitable for a receiver RF front-end) or high output power (for a transmitter front-end). Table 6.3 shows the performance outlook for two well-established III-IV semiconductor technologies (GaAs and InP) in comparison to the silicon-based technologies reviewed above. It can be observed that despite of the fact that SiGe and CMOS have – and will continue to have – cutoff frequencies comparable to GaAs and InP, devices built in the latter technologies will continue to outperform silicon devices in terms of noise and output power performance. Note further that in the case of III-V based devices, which have been used for mm-wave applications for several years, manufacturable solutions are known for the performance expected in 2014.

6.3   60 GHz Front-End Components

As will become evident in Section 6.4, 60 GHz radios can be highly complex integrated systems, and their overall performance depends on a careful balance of different component specifications. Nevertheless, for initial system design and link budget analysis, the front-end components have the most influence and hence it is valuable to examine their expected performance in detail.

Table 6.3 Comparison of expected RF performance of different semiconductor technologies for 2014 according to ITRS 2008

Technology	CMOS HP	SiGe	GaAs Low noise	GaAs Power	InP Low noise	InP Power
Device type	FET	HBT	MHEMT		HEMT	
Gate length/Emitter width [nm]	20	90	50		35*	
Peak f_T [GHz]	440**	395*	350*	–	420	–
Peak f_{MAX} [GHz]	560**	425**	–	325*	–	450*
NF_{min} at 60 GHz [dB]	2.9	1.3	0.6*	–	0.6*	–
P_{out} at 60 GHz [mW/mm]	–	–	–	600*	–	400*

*Manufacturable solutions are known but not implemented.
**Manufacturable solutions are not known.

6.3.1 60 GHz LNAs in SiGe and CMOS

Table 6.4 presents an overview of recently reported mm-wave LNAs in different silicon technologies. The quoted performance metrics usually correspond to measurements performed at ambient temperature and 60 GHz frequency. It can be observed that the transition from 0.13 μm CMOS to 90 nm CMOS offered a performance advantage, but that the reported results from 65 nm and 90 nm CMOS are in general comparable, which is in agreement with the trend predicted by Table 6.1. Another observation that follows from these results is that while the noise performance of LNAs in 65 nm and 90 nm CMOS approaches that of 13 μm SiGe BiCMOS, this is frequently achieved at the expense of higher power consumption. Note that the reported NF performance for different designs is at best 1.5 dB higher than the NF_{min} figure in Tables 6.1 and 6.2. Moreover, the presented data are drawn from on-wafer measured results that do not take into account additional package losses.

When evaluating a 60 GHz LNA design for a standard-compliant system, it is important to take into consideration its performance over a wide frequency range. In general, for both LNAs and PAs, a tradeoff exists between operating bandwidth and power consumption. By using high-impedance inter-stage matching techniques, a higher gain (and potentially lower NF) can be obtained with a lower bias current. A good example of this design approach is the LNA presented in [11], which achieves a very low power consumption of only 4 mW with a 3 dB bandwidth of approximately 4 GHz. In contrast, to achieve wide-band operation often involves the use of relatively low-impedance match techniques or the use of

Table 6.4 Summary of recently reported silicon LNAs operating in the 60 GHz band

Technology	NF [dB]	Gain [dB]	IP$_{1\,dB}$ [dBm]	Power consumption [mW]	Reference
0.13 μm SiGe	5	>12	−12	8.1	[6]
0.13 μm SiGe	4.5	14.7	−20	10.8	[7]
0.13 μm CMOS	8.8	12	–	54	[8]
0.13 μm CMOS	8	25	−22	79	[9]
90 nm CMOS	5.5	14.6	−14.1	24	[10]
90 nm CMOS	6.5	12.2	−7.2	10.5	[11]
90 nm CMOS	4.4	15	−18	4	[12]
65 nm CMOS	6.2	19	−16	35	[13]
65 nm CMOS	5.6	11.5	–	72	[14]
65 nm CMOS	5.9	15	−15.1	31	[15]

several cascaded stages (each optimized at a slightly different frequency). Both of these techniques increase power consumption.

Figure 6.1 shows an example of measured LNA performance; the data corresponds to the SiGe-based design presented in [1]. A measured NF of 5 dB with about ±1 dB variation across frequency and temperature was obtained. As mentioned above, while the absolute noise performance of silicon-based LNAs is not expected to improve significantly beyond currently reported results, the availability of higher cutoff frequencies in both SiGe and CMOS is expected to improve the stability of such designs over frequency and temperature.

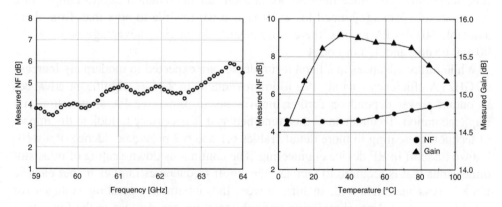

Figure 6.1 Measured results for a SiGe 60 GHz LNA.

6.3.2   60 GHz PAs in SiGe and CMOS

A summary of recently developed PAs for the 60 GHz band is presented in Table 6.5. All of the reported results were measured at ambient temperature. Due to the availability of both a higher supply voltage as well as typically higher achievable voltage gain for a given amplification stage, PAs in SiGe exhibit, in general, higher efficiency and output power than their CMOS counterparts. The differences in performance among reported PA designs in SiGe [16–19] reflect the fact that PAs have a wide design space and face multiple tradeoffs. The PA reported in [18] is a nonlinear design aimed at achieving a high PAE. Non-linear PAs can reduce power consumption in systems where constant-envelope modulations are employed. On the other hand, the PA in [16] was specifically designed to attain a high linearity and achieve an OP1dB as close as possible to the saturated output power, which would be beneficial for modulations with a relatively high PAPR. Moreover, the design in [19] achieves a high saturated output power by using a cascade configuration that enables relatively large output voltage swings from a 4 V supply, at the expense of lower efficiency.

All of the 60 GHz PAs in CMOS reported to date operate in linear mode (class A) or with moderate nonlinearity (class AB). Taking into account that different designs have used different supply voltages (in some cases [10, 24, 13, 26] exceeding the nominal value employed for digital circuits on the same technology), it can be observed that 60 GHz CMOS PAs are capable of delivering 7–10 dBm maximum output power when using a single device at the output and 10–13 dBm when employing two devices at the last amplification stage (with their outputs combined through an on-chip passive component or connected to a differential output). As in the case of LNAs, these numbers should be treated with caution when used for link budget analysis since they do not account for degradation due to temperature variations and package-related losses. It is also worthwhile to note that, from these available 60 GHz PA reports, there is no clear technological advantage between the 90 nm, 65 nm, and 45 nm nodes.

While future advances in silicon technologies are expected to moderately increase 60 GHz PA efficiency (mainly due to improvements in f_{MAX}), significant advances in output power depend on further innovations in mm-wave PA architecture and implementation. To understand this trend better, it is necessary to look at the impact of technology scaling in more detail. Tables 6.1 and 6.2 provide evidence of a well-known tradeoff in RF device engineering. The continuous down-scaling of transistor dimensions that is required to increase the cutoff frequencies implies higher electric fields across materials and, in turn, lowers the maximum operating voltages for reliable operation. Note that during normal operation, the devices in the final stage of the PA will experience voltage swings that exceed the supply voltage. The long-term degradation mechanisms in CMOS PAs operating at mm-wave frequencies are yet to be understood, and for this reason some CMOS PAs have been reported

Table 6.5 Summary of recently reported silicon PAs operating in the 60 GHz band

Technology	Supply voltage [V]	Frequency [GHz]	Gain [dB]	OP1dB [dBm]	Saturated output [dBm]	Peak PAE [%]	Reference
0.25 µm SiGe	3.3	61	18.8	14.5	15.5[1]	19.7	[16]
0.18 µm SiGe	1.8	60	11.5	11.2	15.8[3]	16.8	[17]
0.13 µm SiGe	1.2	58	4.5	–	11.5[1]	20.9	[18]
0.13 µm SiGe	4	60	18	13.1	20[3]	12.7	[19]
0.13 µm CMOS	1.6	60	13.5	7	7.8[1]	3	[20]
90 nm CMOS	1.5	60	5.2	6.4	9.3[1]	7.4	[10]
90 nm CMOS	1.0 / 0.7	60	13.9 / 14.3	10 / 5.2	11[2] / 8.3[2]	8.2 / 6.7	[21]
90 nm CMOS	1.0	60	5.6	9	12.3[2]	8.8	[22]
90 nm CMOS	1.2	63	15	–	10–12.5[3,4]	10–19[4]	[23]
65 nm CMOS	1.2	62	4.5	5.5	9[1]	8	[24]
65 nm CMOS	1.2	60	12.8	1.5	7[1]	–	[13]
65 nm CMOS	1.0	62	15.5	5	11.5[2]	15.2	[25]
45 nm CMOS	1.1	60	6 / 5.6	11 / 8.4	13.8[2] / 10.6[1]	7 / 6.5	[26]

[1]In these designs, the final PA stage employs a single SiGe HBT or CMOS FET device and the output power is measured single-ended.

[2]In these designs, the final PA stage consists of two separate devices. Transmission lines or a transformer are used on chip to combine their output power to a single-ended terminal.

[3]In these designs, the final PA stage consists of two separate devices connected to differential outputs. The reported output power is either the result of off-chip power combining or assumes that the output signals are combined at a differential antenna.

[4]The authors report measurements performed on three different chips.

with measurements under reduced voltage supply conditions [21] to assess their performance under long-term reliability constraints.

The reduction of maximum operating voltage does not necessarily imply a proportional reduction in the amount of power that can be obtained from a semiconductor device in a 60 GHz PA. In principle, a given amount of delivered output power can be maintained with decreased voltage swing as long as the amount of delivered current is increased in the same proportion by using impedance transformation techniques. In a real design the practical extent of this impedance transformation is limited by multiple factors such as the final load impedance presented by the package and antenna, as well as the area and power loss introduced by the passive impedance-transformation network. For these reasons, the development of PA architectures that maintain or increase the available output power at 60 GHz with relatively low supply voltage is an active topic of research.

Since the amount of RF power that can be obtained from a single device is essentially limited by technology, increased output power must instead come from the aggregation of the mm-wave energy originating from different devices. This is achieved through the use of *power combining* techniques. Table 6.5 shows that it is common – and relatively straightforward – to combine the output from two different devices and increase the PA output power by about 3 dB. However, to combine the output of more than two devices requires the implementation of customized complex passive networks. At mm-wave frequencies, these power combiners must be carefully designed leveraging EM simulation and their area overhead may become an important drawback. Figure 6.2 illustrates this area and the tradeoff between complexity and output power with two extreme examples.

The PA illustrated on the left in Figure 6.2 was implemented in SiGe 0.13 μm technology, combines a total of four differential outputs (8 HBT devices in total), and delivers a maximum output power of 23 dBm at 60 GHz using a 4 V supply [27]. A different power combiner design in 90 nm CMOS [28] combines four separate single-ended CMOS amplifiers and attains a maximum 60 GHz output power of 14.2 dBm from a 1 V supply. Both designs achieve a higher saturated output compared

Multi-stage SiGe PA with transformer-based power
combining: 1.9 mm × 1.8 mm , 23 dBm Psat

Single-stage CMOS PA:
450 um × 600 um
9 dBm Psat

Figure 6.2 Two examples of 60 GHz PA designs illustrating a tradeoff between silicon area and output power.

to any other PA from the same technology in Table 6.5. However, both use more than 1 mm^2 of silicon (which is comparable to a digital baseband IC or some of its portions) and achieve a maximum PAE of only 6%.

6.3.3 Process Variability in Silicon Millimeter-Wave Designs

A separate but nonetheless important consideration for both PAs and LNAs is the process variability. For analog and mixed-signal circuits with moderate to high complexity (such as ADCs, op-amp based signal processing units, or complex supply regulators) the increased device variability introduced by device scaling is a growing concern. The development of techniques to address this variability is an active topic of investigation. Millimeter-wave front-end circuits are different in that their complexity lies in the precise (in terms of layout, and EM modeling) integration of active components with custom-designed passive elements rather than in the large-scale interaction of multiple transistors. Given the relatively low device count, and because the RF properties (power gain, noise, and parasitics) for a given bias current are very stable across process [5], mm-wave amplifiers can show a remarkable performance uniformity against process variations even when realized using simple constant-current biasing techniques. As an example, Figure 6.3 shows

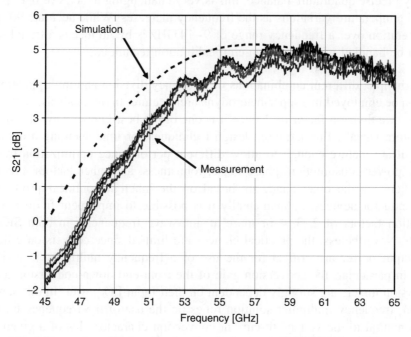

Figure 6.3 Measured small-signal gain of a single-stage CMOS PA for different wafers.

the measured small-signal gain of a single-stage 65 nm CMOS PA IC [24] sampled from different wafers showing a variability of about 0.5 dB in the 60 GHz band.

6.4 Frequency Synthesis and Radio Architectures

The overall architecture of a 60 GHz radio is closely related to its frequency planning and frequency synthesis strategy; both architecture and frequency planning will be discussed in this section. Figure 6.4 presents a generic super-heterodyne architecture for a 60 GHz transceiver chipset. A two-step conversion architecture for 60 GHz applications has two main advantages from a local oscillator (LO) signal generation viewpoint:

1. The VCO can operate at frequencies below 30 GHz where wider frequency tuning range and phase noise can be obtained, especially considering process and temperature variations. This is particularly important for applications that target use of all of the frequency channels and at least some of the complex modulation schemes of standards such as IEEE 802.15.3c.
2. The quadrature of the up-conversion and down-conversion signals is introduced at the first up/down-conversion step at a frequency below 15 GHz and through the use of a divider. SC and OFDM signaling schemes at Gbps data rates require very precise quadrature balance; this is very challenging to achieve over process and temperature variations as the frequency increases. While accurate 90° phase generation over a frequency range of 9–14 GHz is by no means a trivial task, it can be handled with known on-chip calibration techniques.

In principle, different combinations of frequency division and multiplication factors can be employed in a super-heterodyne architecture. In our analysis, we focus on the use of a division factor of 2 since it is one of the best-known ways of obtaining quadrature signals. Furthermore, design techniques for implementing a divide-by-2 circuit are mature enough to allow different performance optimization tradeoffs among power consumption, phase noise, robustness, and other factors. Frequency multipliers, on the other hand, are based on the inherent nonlinear properties of semiconductor devices. Fundamentally it is possible to implement frequency multiplication factors of 2, 3, 4 or more at mm-wave frequencies in both SiGe and CMOS. Nevertheless, the practical choices are limited since conversion efficiency and output power are critical for the use of a frequency multiplier in a 60 GHz radio. In particular, the conversion gain of the front-end down-conversion and up-conversion mixers are strongly dependent on their input LO power. In this sense, an efficient frequency multiplier solution relies on the use of the frequency harmonic that is natural to the voltage-to-current conversion characteristics of a given semiconductor device. This consideration leads to the conclusion that, in general, the

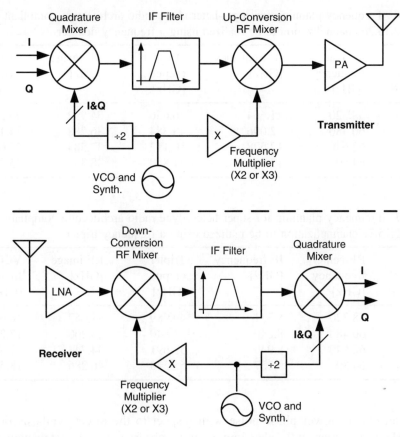

Figure 6.4 Generic super-heterodyne receiver and transmitter radio architecture using a frequency multiplier.

use of a frequency tripler is more power efficient in SiGe and a frequency doubler more power efficient in CMOS. An example of the implementation of a 60 GHz transmitter and receiver in SiGe that uses a frequency tripler is presented in [7].

Having outlined the design considerations in favor of the use of a divide-by-2 circuit and a multiplication factor of either 2 or 3, Tables 6.6 and 6.7 outline the full frequency planning of a super-heterodyne radio to comply with the channel plan of the IEEE 802.15.13c standard, as described in Chapter 1. The use of multiplication by 2 results in higher VCO and IF frequencies but makes the image frequency easier to filter.

Table 6.8 presents a summary of reported frequency synthesis solutions for 60 GHz radios. Although a frequency multiplier degrades the LO phase noise with respect to that of the source VCO, solutions that employ a multiplication factor

Table 6.6 Frequency planning for super-heterodyne radio architecture compliant with
IEEE 802.15.3c channelization to be realized using a frequency doubler

IEEE 802.15.3c Channel #	RF center frequency [GHz]	IF frequency [GHz]	Doubler output [GHz]	RF image [GHz]	VCO frequency [GHz]
1	58.320	11.664	46.656	34.992	23.328
2	60.480	12.096	48.384	36.288	24.192
3	62.640	12.528	50.112	37.584	25.056
4	64.80	12.960	51.840	38.880	25.920

Table 6.7 Frequency planning for super-heterodyne radio architecture compliant with
IEEE 802.15.3c channelization to be realized using a frequency tripler

IEEE 802.15.3c Channel #	RF center frequency [GHz]	IF frequency [GHz]	Tripler output [GHz]	RF image [GHz]	VCO frequency [GHz]
1	58.320	8.331	49.989	41.657	16.663
2	60.480	8.640	51.840	43.200	17.280
3	62.640	8.949	53.691	44.743	17.897
4	64.80	9.257	55.543	46.286	18.514

achieve an overall lower phase noise with respect to the direct synthesis of a 50 or 60 GHz LO carrier. It is also important to observe that two-step conversion solutions naturally enable coverage of a wider range of frequencies as compared to direct alternatives.

6.5 Radio–Baseband Interface

6.5.1 ADCs and DACs for Wide Bandwidth Signals

With respect to previous wireless technologies, such as 802.11a/b/g/n, 60 GHz systems augment the employed channel bandwidth by a factor of 10 or more. This increase, which is the principal factor in enabling data rates in excess of 1 Gb/s, demands a proportional increase in the sampling rate of analog-to-digital and digital-to-analog interfaces in the system. To process the channel bandwidth of signaling schemes in the 802.15.3c standard (approx. 850 MHz at baseband), a sampling rate of at least 1.7 Gsps is necessary. Nevertheless, proper symbol synchronization requires some degree of oversampling, and sampling rates in excess of 2.5 Gsps will be required in most practical 60 GHz systems.

Table 6.8 Summary of synthesis solutions for 60 GHz transceiver applications

Reference	Type	Frequency [GHz]	In-band phase noise [dBc/Hz]	Phase noise @ 1 MHz [dBc/Hz]	Power consumption [mW]	Technology
[29]	Fundamental	55.0–58.0	−58.0	−72.0	650	SiGe:C
[7]	Tripled	50.0–59.0	−83.5	−87.5	75+86*	0.13 μm SiGe
[30]	Fundamental	46.0–51.0	−63.5	−72.0	57	0.13 μm CMOS
[31]	Fundamental	58.0–60.0	−	−85.1	80	90 nm CMOS
[32]	Tripled	46.0–54.0	−80.5	−90.5	145+113*	0.13 μm SiGe
[33]	Doubled**	64.3–66.2	−	−84.0	72	0.13 μm CMOS

*Includes power consumption of tripler and output buffers.
**Through a push-push VCO instead of a frequency doubler.

To understand the challenge involved in attaining these conversion speeds and – once again the impact of technology scaling on the expected performance, we examine the recent trends in silicon integrated ADCs and DACs. Table 6.9 presents a summary of ADCs reported in the last 5 years which feature sample frequencies greater than 400 MHz [34–53]. The designs are grouped by technology node to support analysis of performance trends. Due to the high speed switching nature of data conversion operations (similar to those in digital circuits), CMOS is the clear technology of choice for data converters within this range of sampling frequencies.

In addition to the designed number of bits and maximum sampling frequency, Table 6.9 includes the effective number of bits (ENOB) and effective resolution bandwidth (ERBW). The ENOB is calculated from the measured signal-to-noise-and-distortion ratio (SNDR) that corresponds to an input signal with a frequency equal to the ERBW (usually the Nyquist frequency). In other words, the ADC has at least x ENOB over the signal bandwidth ERBW. To compare ADC conversion efficiencies, the following figure of merit (FOM) is generally employed in the literature:

$$FOM = \frac{\text{Power consumption}}{(2 \cdot \text{ERBW}) \cdot (2^{\text{ENOB}})}[\text{pJ/conv}]. \tag{6.1}$$

ADCs with high sampling rates have an ENOB performance that tends to decrease with input frequency due to multiple design impairments such as clock jitter, mismatch between components, and nonlinearities. Robust ADCs maintain an

Table 6.9 Summary of ADCs for wide-bandwidth signal processing applications

Process [nm]	Design bits	ENOB	Max. sampling freq. [Gsps]	ERBW [GHz]	Power consumption [mW]	FOM [pJ/conv.]	Architecture	Reference
90	5	3.6	3.5	1.00	227	9.18	Flash	[34]
90	7	5.2	0.8	0.30	120	5.48	Folding-interpolation	[35]
90	6	5.1	10.7	5.00	1600	4.70	Time-interleaved pipeline	[36]
90	6	5.3	1.0	0.50	55	1.37	Two-step sub-ranging	[37]
90	4	3.7	1.3	0.63	2.5	0.16	Modified flash	[38]
90	6	4.9	0.6	0.33	10	0.52	Time-interleaved	[39]
90	11	8.7	0.8	0.40	350	1.07	Time-interleaved	[40]
90	9	7.0	0.4	0.16	139	3.35	Pipeline	[41]
90	7	6.0	1.1	0.30	46	1.18	Time-interleaved pipeline	[42]
90	6	4.9	3.5	1.75	98	0.94	Flash	[43]
90	5	4.3	1.8	0.88	2.2	0.06	Folding flash	[44]
90	5	4.6	1.8	0.88	7.6	0.18	Flash	[45]
90	8	6.9	1.3	0.63	207	1.44	Folding flash	[46]
90	6	5.3	2.7	1.35	50	0.47	Flash	[47]
65	5	2.9	0.5	0.25	7.5	2.06	Time-interleaved SAR	[48]
65	6	5.2	0.8	0.40	12	0.41	Flash	[49]
65	6	5.2	5.0	2.50	320	1.75	Flash	[50]
65	4.5	3.8	7.5	3.75	52	0.51	Flash	[51]
45	7	5.4	2.5	1.25	52	0.51	Time-interleaved	[52]
45	6	5.5	1.2	0.60	28.5	0.52	Flash	[53]

ENOB closer to the designed number of bits over an ERBW closer to the Nyquist frequency; this is in general the desired ADC performance for wide bandwidth wireless applications.

It is important to note that in most ADC publications the reported power consumption corresponds exclusively to the *ADC core*; the unit that actually performs the conversion. In practice, high-speed ADCs require additional components such as clock buffers, drivers for analog and digital signals connected externally through 50 Ω interfaces, and voltage regulators. The performance of this additional circuitry is vital not only to ADC operation but also to maintain the required ENOB over

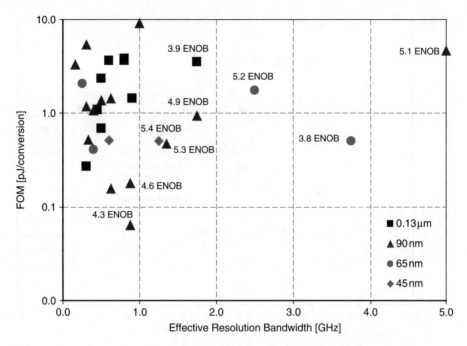

Figure 6.5   ADC conversion efficiency versus ERBW for different technology nodes.

voltage, process and temperature variations. The power and silicon area associated with these components is often comparable to that of the ADC core itself.

In order to give a better understanding of the trends in ADC efficiency for different technology nodes, Figures 6.5 and 6.6 present the ADC FOM as a function of ENOB and ERBW for the designs in Table 6.9. Additional ADCs in a 0.13 μm CMOS process are also plotted in these graphs to provide a technology-scaling perspective. At low to moderate ENOB (less than 5.5 bits) and ERBW less than 1 GHz, there are several examples of ADCs achieving an FOM<1 pJ/conv. Nevertheless, as both speed and resolution increase, efficiency decreases exponentially.

The transition from a 0.13 μm to a 90 nm process combined with innovations in ADC architecture and calibration algorithms, resulted in improved conversion efficiency versus ERBW, as can be observed in Figure 6.5. The availability of a 65 nm technology enabled ADCs with higher sampling frequencies [50, 51], but so far has not demonstrated an increased efficiency for a given ERBW. When comparing conversion efficiency with respect to ENOB (Figure 6.6) it can be observed that both 90 nm and 65 nm technologies offer an advantage with respect to 0.13 μm for ADCs with relatively low resolution but not for ENOB greater than 5.5 dB. These observations indicate that as technology scaling progresses, it is easier to achieve faster conversion speeds (due to higher f_T and f_{MAX});

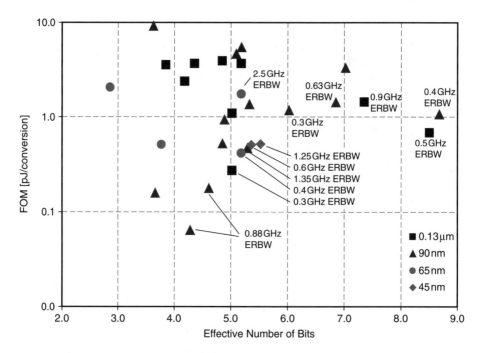

Figure 6.6   ADC power consumption versus ENOB for different technology nodes.

nevertheless the associated reduction in voltage supply and increased device mismatch (detrimental to comparators and other ADC circuits) make it difficult to sustain the ENOB performance and/or to increase the overall conversion efficiency.

In this context, it is clear that continuous innovation in ADC design is a key to making power consumption acceptable in 60 GHz systems. For example, the ADC reported in [44] (2.5 mW for about 600 MHz ERBW and 4 ENOB) has the best conversion efficiency (0.16 pJ/conv) among the published ADCs at these speeds and is a major advance in the state of the art. For a 2 Gbps QPSK OFDM system, the ADC requirements are approximately 5 ENOB and 1 GHz of ERBW. So far, only two of the reported ADCs meet these specifications [47, 52] and they do so at an FOM of 0.5 pJ/conv. It is interesting to observe that these ADCs were implemented in significantly different technologies (90 nm and 45 nm) but achieve almost identical performance by applying different calibration and architecture concepts. Further design innovations and possibly the use of a 32 nm process will be required to maintain this performance with the required robustness to process and temperature variations required for a commercial 60 GHz system. For 3 Gbps or higher using 16-QAM OFDM over the same 1 GHz bandwidth, the required resolution would be about 8 bits (based on extrapolation from 802.11a designs).

Table 6.10 Summary of current state of the art in high-speed DACs

Technology	Design bits	Max. sample rate [Gs/s]	Measured SFDR at output frequency	Power [W]	Reference
0.25 µm BiCMOS	15	1.2	63 dB at 1.2 GHz	6	[54]
0.35 µm CMOS	10	1.0	61 at 490 MHz	0.11	[55]
0.18 µm CMOS	14	1.4	67 dB at 260 MHz	0.4	[56]
65 nm CMOS	12	2.9	60 at 550 MHz	0.19	[57]

If an efficiency of 0.8 pJ/conv is achieved at future levels of performance, the corresponding ADC would consume approximately 400 mW.

A smaller number of examples are found in the literature for high speed DACs [54–57]. Some of the relevant designs are listed in Table 6.10. The required characteristics for a Gbps OFDM design in a 60 GHz system would be approximately 50 dB of spurious-free-dynamic range (SFDR, analogous performance metric to SNDR in ADCs) and a sample frequency greater than 2 GHz. The DAC reported in [57] meets these specifications with a power consumption of 180 mW. The performance trends previously discussed for ADCs are expected to be similar in the case of DACs.

6.5.2 Modulators, Demodulators and Analog Signal Processors for Gbps Applications

As can be understood from the previous subsection, the resources (design complexity, power, area) demanded by a high-speed digital modulation engine along with an ADC/DAC operating at GHz clock sampling rates may become an important bottleneck for the incorporation of these high-speed radios into portable devices. This has motivated the development of integrated modulators and demodulators for relatively simple modulation schemes, trading off spectral efficiency for reduced system complexity.

Before the availability of silicon-integrated mm-wave integrated circuits, 60 GHz communication systems based on monolithic microwave integrated circuits were already employing modulators and demodulators for OOK [58] and BSK [59]. This trend continues in the most recent examples of CMOS 60 GHz transceivers which employ SC modulation schemes and incorporate several signal-processing functions in the analog and/or mixed-signal domains. Two 60 GHz transceivers in 90 nm CMOS integrate OOK modulator and demodulator [60, 61], while a reported transceiver in 65 nm CMOS supports BPSK [62]. Furthermore, the transceiver in [63] includes a QAM modulator in the transmitter and an equalizer in the receiver path. MSK signaling offers the advantages of being a constant-envelope modulation

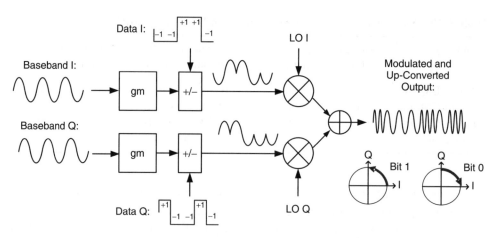

Figure 6.7   Block diagram of a compact integrated modulator.

with higher spectral efficiency than QPSK, BPSK and OOK. Moreover, as discussed in Chapter 1, it is employed as common-mode modulation in the IEEE 802.15.3c standard. A mixed-signal baseband integrated circuit (intended for integration in a 60 GHz receiver) for the coherent detection of MSK modulation is presented in [64] and achieves up to 1 Gbps with only 55 mW.

As an example of an integrated modulator for Gbps data rates, an MSK modulator that has been integration with a sliding-IF super-heterodyne 60 GHz transmitter [65] is now described. Figure 6.7 is a conceptual block diagram of this modulator. Transconductors (gm) convert quadrature baseband sinusoidal signals into current. The digital data enters the modulator in two branches (I and Q) and controls the polarity of the current. Both current signals are multiplied by an LO signal and then combined. This operation completes the modulation procedure and up-converts the resultant signal to a desired RF frequency which, in a super-heterodyne radio, can be the sliding-IF frequency.

MSK modulation is achieved by applying quadrature sinusoids to the baseband I and Q inputs, along with appropriately coded and synchronized binary polarity control signals on the Data I and Data Q lines. The polarities of the cosine and sine pulses change at their crossover points to encode a bit stream onto the frequency of the output signal. In this example encoding, an information data bit of 1 generates a positive frequency and a data bit of 0 generates a negative frequency. The frequency changes polarity at I/Q quadrant boundaries as shown in Figure 6.7. The resulting MSK modulated signal has the attractive properties of constant envelope and good spectral efficiency. As a specific example, to generate a 2 Gbps MSK data stream, quadrature I/Q tones with a frequency of 500 MHz are input to the MSK modulator baseband I/Q ports while a 2 Gbps data stream is split and encoded into two 1 Gbps data streams to control the polarity of the sinusoids. Since, as described above, the MSK encoding produces a positive frequency shift when transmitting a '1' bit and a negative shift when transmitting a '0' bit, this signal can be demodulated by a conventional FM discriminator. Circuit implementation details for both the modulator and demodulator are provided in [66], including a demonstration of their operation up to 2 Gbps.

References

[1] Reynolds, S., Floyd, B., Pfeiffer, U. and Zwick, T. (2004) 60 GHz transceiver circuits in SiGe bipolar technology. *IEEE International Solid-State Circuits Conference*, pp. 442–538, February.

[2] Niknejad, A.M. and Hashemi, H. (eds) (2008) *mm-Wave Silicon Technology: 60 GHz and Beyond*. New York: Springer.

[3] http://www.itrs.net/

[4] Jhon, H.-S. et al. (2009) fMAX improvement by controlling extrinsic parasitics in circuit-level MOS transistor. *IEEE Electron Device Letters*, **30**(12), 1323–1325.

[5] Dickson, T.O. et al. (2006) The invariance of characteristic current densities in nanoscale MOSFETs and its impact on algorithmic design methodologies and design porting of Si(Ge) (Bi)CMOS high-speed building blocks. *IEEE Journal of Solid-State Circuits*, **45**(8), 1830–1845.

[6] Alvarado, J., Kornegay, K.T., Dawn, D., Pinel, S. and Laskar, J. (2007) 60-GHz LNA using a hybrid transmission line and conductive path to ground technique in silicon. *IEEE Radio Frequency Integrated Circuits Symposium*, pp. 685–688, June.

[7] Reynolds, S., Floyd, B., Pfeiffer, U., Beukema, T., Grzyb, J., Haymes, C., Gaucher, B. and Soyuer, M. (2006) A silicon 60 GHz receiver and transmitter chipset for broadband communications. *IEEE Journal of Solid-State Circuits*, **41**(12), 2820–2831.

[8] Doan, C.H., Emami, S., Niknejad, A.M. and Brodersen, R.W. (2005) Millimeter-wave CMOS design. *IEEE Journal of Solid-State Circuits*, **40**(1), 144–155.

[9] Lo, C.-M. et al. (2006) A miniature V-band 3-stage cascode LNA in 0.13 um CMOS. *IEEE International Solid-State Circuits Conference*, pp. 1254–1263, February.

[10] Yao, T., Gordon, M.Q., Tang, K.K.W., Yau, K.H.K., Yang, M.-T., Schvan, P. and Voinigescu, S.P. (2007) Algorithmic design of CMOS LNAs and PAs for 60-GHz radio. *IEEE Journal of Solid-State Circuits*, **42**(5), 1044–1057.

[11] Heydari, B., Bohsali, M., Adabi, A. and Niknejad, A.M. (2007) Millimeter-wave devices and circuit blocks up to 104 GHz in 90 nm CMOS. *IEEE Journal of Solid-State Circuits*, **42**(12), 2893–2903.

[12] Cohen, E., Ravid, S. and Ritter, D. (2008) An ultra low power LNA with 15 B gain and 4.4 dB N in 90 nm CMOS process for 60 GHz phase array radio. *IEEE Radio Frequency Integrated Circuits Symposium*, pp. 61–64, June.

[13] Weyers, C. et al. (2008) A 22.3 dB voltage gain 6.1 dB NF 60 GHz LNA in 65 nm CMOS with differential output. *IEEE International Solid-State Circuits Conference*, pp. 192–193, February.

[14] Varonen, M., Kärkkäinen, M., Kantanen, M. and Halonen, K.A.I. (2008) *Millimeter-wave integrated circuits in 65-nm CMOS*. *IEEE Journal of Solid-State Circuits*, *43 (9),1991–2002*.

[15] Natarajan, A. et al. (2008) A 60 GHz variable-gain LNA in 65 nm CMOS. *IEEE Asian Solid-State Circuits Conference*, pp. 117–120, November.

[16] Do, V.-H., Subramanian, V., Keusgen, W. and Boeck, G. (2008) A 60 GHz SiGe-HBT power amplifier with 20% PAE at 15 dBm output power. *IEEE Microwave and Wireless Component Letters*, **18**(3), 209–211.

[17] Wang, C. et al. (2006) A 60 GHz transmitter with integrated antenna in a 0.18 μm SiGe BiC-MOS technology. *IEEE International Solid-State Circuits Conference*, pp. 186–187, February.

[18] Valdes-Garcia, A. et al. (2007) A 60 G GHz class-E power amplifier in SiGe. *IEEE Asian Solid-State Circuits Conference*, pp. 199–202, November.

[19] Pfeiffer U.R. and Goren, D. (2007) A 20 dBm fully-integrated 60 GHz SiGe power amplifier with automatic level control. *IEEE Journal of Solid-State Circuits*, **42**(7), 1455–1463.

[20] Wicks, B. et al. (2008) A 60-GHz fully-integrated Doherty power amplifier based on 0.13-um CMOS process. *IEEE Radio Frequency Integrated Circuits Symposium*, pp. 65–68, June.

[21] Tanomura, M. et. al. (2006) TX and RX front-ends for 60 GHz band in 90 nm standard bulk CMOS. *IEEE International Solid-State Circuits Conference*, pp. 558–559, February.

[22] Chowdhury, D., Reynaert, P. and Niknejad, A.M. (2009) Design considerations for 60 GHz transformer-coupled CMOS power amplifiers. *IEEE Journal of Solid-State Circuits*, **44**(10), 2733–2744.

[23] LaRocca, T., Liu, J.Y.-C. and Chang, M.-C.F. (2009) 60 GHz CMOS amplifiers using transformer-coupling and artificial dielectric differential transmission lines for compact design. *IEEE Journal of Solid-State Circuits*, **44**(5), 1425–1435.

[24] Valdes-Garcia A. et. al. (2008) 60 GHz transmitter circuits in 65 nm CMOS. *IEEE Radio Frequency Integrated Circuits Symposium*, pp. 641–644, June.

[25] Chan, W.L. and Long, J.R. (2010) A 58–65 GHz neutralized CMOS power amplifier with PAE above 10% at 1-V supply. *IEEE Journal of Solid-State Circuits*, **45**(3), 554–564.

[26] Raczkowski, K., Thijs, S., De Raedt, W., Nauwelaers, B. and Wambacq, P. (2009) 50-to-67 GHz ESD-Protected Power Amplifiers in Digital 45 nm LP CMOS. *IEEE International Solid-State Circuits Conference*, pp. 382–383, February.

[27] Pfeiffer U.R. and Goren, D. (2007) A 23 dBm 60 GHz distributed active transformer in a silicon process technology. *IEEE Transactions on Microwave Theory and Techniques*, **55**(5), 857–865, May.

[28] Bohsali, M. and Niknejad, A.M. (2009) Current combining 60 GHz CMOS power amplifiers. *IEEE Radio Frequency Integrated Circuits Symposium*, pp. 31–33, June.

[29] Winkler, W., Borngräber, J., Heinemann, B. and Herzel, F. (2005) A fully integrated BiCMOS PLL for 60 GHz wireless applications. *IEEE International Solid-State Circuits Conference*, pp. 406–407, February.

[30] Cao, C., Ding, Y. and O, K.K. (2006) A 50-GHz PLL in 130-nm CMOS. *IEEE Custom Integrated Circuits Conference*, pp. 21–24, September.

[31] Lee, C. and Liu, S.-I. (2007) A 58-to-60.4 GHz frequency synthesizer in 90 nm CMOS. *IEEE International Solid-State Circuits Conference*, pp. 196–197, February.

[32] Floyd, B. (2007) A 15 to 18-GHZ programmable sub-integer frequency synthesizer for a 60-GHz transceiver. *IEEE Radio Frequency Integrated Circuits Symposium*, pp. 529–532, June.

[33] Tsai, K.-H. (2008) A digitally calibrated 64.3-66.2 GHz phase-locked loop. *IEEE Radio Frequency Integrated Circuits Symposium*, pp. 307–311, June.

[34] Park, S. Palaskas, Y., Ravi, A., Bishop, R.E. and Flynn, M.P. (2006) A 3.5 GS/s 5-b flash ADC in 90 nm CMOS. *IEEE Custom Integrated Circuits Conference*, pp. 489–492, September.

[35] Makigawa, K., Ono, K., Ohkawa, T., Matsuura, K. and Segami, M. (2006) A 7 bit 800 Msps 120 mW folding and interpolation ADC using a mixed-averaging scheme. *IEEE Symposium on VLSI Circuits*, pp. 15–17, June.

[36] Nazemi, A. et al. (2008) A 10.3 GS/s 6 bit (5.1 ENOB at Nyquist) time-interleaved/pipelined ADC using open-loop amplifiers and digital calibration in 90 nm CMOS. *IEEE Symposium on VLSI Circuits*, pp. 18–19, June.

[37] Figueiredo, P.M. et al. (2006) A 90 nm CMOS 1.2 V 6 b 1 GS/s two-step subranging ADC. *IEEE International Solid-State Circuits Conference*, pp. 2320–2321, February.

[38] Der Plas, G.V., Decoutere, S. and S. Donnay, S. (2006) A 0.16 pJ/conversion-step 2.5 mW 1.25 GS/s 4b ADC in a 90 nm digital CMOS process. *IEEE International Solid-State Circuits Conference*, pp. 2310–2311, February.

[39] Draxelmayr, D. (2004) A 6 b 600 MHz 10 mW ADC array in digital 90 nm CMOS. *IEEE International Solid-State Circuits Conference*, pp. 264–255, February.

[40] Hsu, C.-C., Huang, F.-C., Shih, C.-Y., Huang, C.-C., Lin, Y.-H., Lee C.-C. and Razavi, B. (2007) An 11 b 800 MS/s time-interleaved ADC with digital background calibration. *IEEE International Solid-State Circuits Conference*, February.

[41] Peach, C.T., Ravi, A., Bishop, R., Soumyanath, K. and Allstot, D.J. (2005) A 9-b 400 Msample/s pipelined analog-to-digital converter in 90 nm CMOS. *IEEE European Solid-State Circuits Conference*, pp. 535–538, September.

[42] Hsu, C-C. et al. (2007) A 7 b 1.1 GS/s reconfigurable time-interleaved ADC in 90 nm CMOS. *IEEE Symposium on VLSI Circuits*, pp. 66–67, June.

[43] Deguchi, K. (2007) A 6-bit 3.5-GS/s 0.9-V 98-mW flash ADC in 90 nm CMOS. *IEEE Symposium on VLSI Circuits*, pp. 64–65, June.

[44] Verbruggen, B. et al. (2009) A 2.2 mW 5 b 1.75 GS/s folding flash ADC in 90 nm digital CMOS. *IEEE Journal of Solid-State Circuits*, **44**(3), 874–882.

[45] Verbruggen, B. et al. (2008) A 7.6 mW 1.75 GS/s 5 bit flash A/D converter in 90 nm digital CMOS. *IEEE Symposium VLSI Circuits*, pp. 14–15, June.

[46] Yu, H. and Chang, M.-C.F. (2008) A 1-V 1.25-GS/s 8-bit self-calibrated flash ADC in 90-m digital CMOS. *IEEE Transactions on Circuits and Systems II*, **55**(7), 668–672.

[47] Nakajima, Y., Sakaguchi, A., Ohkido, T., Kato, N., Matsumoto, T. and Yotsuyanagi, M. (2010) A background self-calibrated 6 b 2.7 GS/s ADC with cascade-calibrated folding-interpolating architecture. *IEEE Journal of Solid-State Circuits*, **45**(4), 707–718.

[48] Ginsburg, B.P. and Chandrakasan, A. (2006) A 500 MS/s 5 b ADC in 65 nm CMOS. *IEEE Symposium on VLSI Circuits*, pp. 140–141, June.

[49] Chen, C.-Y. (2008) A low power 6-bit flash ADC with reference voltage and common-mode calibration. *IEEE Symposium on VLSI Circuits*, pp. 12–13, June.

[50] Choi, M. (2008) A 6-bit 5-GSample/s Nyquist A/D converter in 65 nm CMOS. *IEEE Symposium on VLSI Circuits*, pp. 16–17, June.

[51] Chung, H., Rylyakov, A., Deniz, Z.T., Bulzacchelli, J., Wei, G.-Y. and Friedman, D. (2009) A 7.5-GS/s 3.8-ENOB 52-mW flash ADC with clock duty cycle control in 65 nm CMOS. *IEEE Symposium on VLSI Circuits*, pp. 268–269, June.

[52] Alpman, E., Lakdawala, H., Carley, L.R. and Soumyanath, K. (2009) A 1.1 V 50 mW 2.5 GS/s 7 b time-interleaved C-2C SAR ADC in 45 nm LP digital CMOS. *IEEE International Solid-State Circuits Conference*, pp. 76–77, February.

[53] Veldhorst, P. et al. (2009) A 0.45 pJ/conv-step 1.2 Gs/s 6 b full-Nyquist non-calibrated flash ADC in 45 nm CMOS and its scaling behavior. *IEEE European Solid-State Circuits Conference*, pp. 464–467, September.

[54] Jewett, B., Liu, J. and Poulton, K. (2005) A 1.2 GS/s 15 b DAC for precision signal generation. *IEEE International Solid-State Circuits Conference*, pp. 110–111, February.

[55] Van de Bosch, A., Borremans, M.A.F., Steyaert, M.S.J. and Sansen, W. (2001) A 10-bit 1-GSample/s Nyquist current-steering CMOS D/A converter. *IEEE Journal of Solid-State Circuits*, **36**(3), 315–324.

[56] Schafferer, B. and Adams, R. (2004) A 3V CMOS 400 mW 14 b 1.4 Gs/s DAC for multi-carrier applications. *IEEE International Solid-State Circuits Conference*, pp. 360–361, February.

[57] Lin, C.-H. et al. (2009) A 12 bit 2.9 GS/s DAC with IM3 < − 60 dBc beyond 1 GHz in 65 nm CMOS. *IEEE Journal of Solid-State Circuits*, **44**(12), 3285–3293.

[58] Ohata, K. et al. (2002) Wireless 1.25 Gb/s transceiver module at 60 GHz band. *IEEE International Solid-State Circuits Conference*, pp. 236–237, February.

[59] Sarkar, S., Yeh, D.A., Pinel, S. and Laskar, J. (2006) 60-GHz direct-conversion gigabit modulator/demodulator on liquid-crystal polymer. *IEEE Transactions on Microwave Theory and Techniques*, **54**(3), 1245–1252.

[60] Juntunen, E. et al. (2010) A 60-GHz 38-pJ-bit 3.5-Gb/s 90-nm CMOS OOK digital radio. *IEEE Transactions on Microwave Theory and Technique*. **52**(2), 348–355.

[61] Lee, J., Chen, Y. and Huang, Y. (2010) A low-power low-cost fully-integrated 60-GHz transceiver system with OOK modulation and on-board antenna assembly. *IEEE Journal of Solid-State Circuits*, **45**(2), 264–275.

[62] Tomkins, A., Aroca, R.A., Yamamoto, T., Nicolson, S.T., Doi, Y. and Voinigescu, S.P. (2009) A zero-IF 60 GHz 65 nm CMOS transceiver with direct BPSK modulation demonstrating up to 6 Gb-s data rates over a 2 m wireless link. *IEEE Journal of Solid-State Circuits*, **44**(8), 2085–2099.

[63] Marcu, C. et al. (2009) A 90 nm CMOS low-power 60 GHz transceiver with integrated baseband circuitry. *IEEE Journal of Solid-State Circuits*, **44**(12), 3434–3447.

[64] Sobel, D.A. and Brodersen, R.W. (2009) A 1 Gb/s mixed-signal baseband analog front-end for a 60 GHz wireless receiver. *IEEE Journal of Solid-State Circuits*, **44**(4), 1281–1289.

[65] Reynolds, S., Valdes-Garcia, A., Floyd, B., Gaucher, B., Liu, D. and Hivik, N. (2007) Second generation transceiver chipset supporting multiple modulations at Gb/s data rates. *IEEE Bipolar/BiCMOS Circuits and Technology Meeting*, pp. 192–197, October.

[66] Valdes-Garcia, A., Reynolds, S. and Beukema, T. (2007) Multi-mode modulator and frequency demodulator circuits for Gb/s data rate 60 GHz wireless transceivers. *IEEE Custom Integrated Circuits Conference*, pp. 639–642, September.

7

Hardware Implementation for Single Carrier Systems

Yasunao Katayama

7.1 Introduction

Although various wireless standards define specifications for the air interface rather than from the transmitter side of the link, the overall system performance is often determined by the receiver-side design.[1] In 60 GHz systems, this situation is becoming even more apparent due to various implementation challenges. Both carrier frequency and data transmission rates are an order of magnitude higher than in conventional wireless systems such as 5 GHz wireless local area networks. As a result, while major challenges in conventional wireless systems exist in recovering the channel disturbances, those in 60 GHz involve radio frequency (RF) as well as some baseband-related circuits (data converters, etc.) that could behave significantly different from ideal models in popular simulation environments such as MATLAB.

In this chapter we will discuss single carrier (SC) system implementation examples, starting with two proof-of-concept implementations equipped with full digital baseband, one with non-coherent detection and the other with differentially coherent detection. These systems were designed as early-phase prototyping; nevertheless, they provide significant technical insight for the design of multi-Gbps wireless systems with digital baseband techniques. The modulation formats used in the non-coherent demonstration system described here have been adopted as

[1]The author would like to acknowledge colleagues in the IBM Research mmWave team for their collaboration in addressing the extremely challenging issues discussed in this chapter.

60 GHz Technology for Gbps WLAN and WPAN: From Theory to Practice
Edited by Su-Khiong (SK) Yong, Pengfei Xia and Alberto Valdes Garcia
© 2011 John Wiley & Sons, Ltd

common/mandatory modes in IEEE 802.15.3c. However, the frame formats employed, including both random error and packet loss forward error correction (FEC) recovery, were custom implementations (non-standard) designed to compensate for non-idealities from early RF prototypes. Then, after presenting the test and evaluation results, we will discuss how to implement more advanced SC systems that can comply with a given standard such as IEEE 802.15.3c.

We assume that, with SC modulation schemes, a low-power and low-cost 60 GHz system implementation is feasible under line-of-sight (LOS) channel environments, when non-coherent detection is used, and that high-performance implementation is also an option with additional signal processing, when coherent detection is exploited.

All the implementation examples discussed here were conducted with real-time field-programmable gate array (FPGA) based platforms. Compared with non-real-time software baseband implementations, more realistic designs can be tested and verified with various real-time feedback loops. Compared with application-specific integrated circuit (ASIC) designs, the FPGA-based approaches help to reduce design risks given unknown channel and RF characteristics and rapidly changing standards. The design can be easily converted to ASIC for final products, resulting in efficient design in terms of cost and power consumption. Further details on non-coherent and differentially coherent system implementations can be found in [1, 2] while further details on advanced SC systems with coherent detection can be found in [3–5].

This chapter is organized as follows. Section 7.2 briefly describes the advantages and challenges of SC system, in particular highlighting implementation aspects. Sections 7.3 and 7.4 present system design with non-coherent detection and differentially coherent detection, respectively. After test and evaluation results of these systems in Section 7.5, we will discuss an advanced SC system with per-packet coherent detection that can comply with the IEEE 802.15.3c SC standard in Section 7.6. Our conclusions are presented in Section 7.7.

7.2 Advantages and Challenges of SC Systems

Assuming more than 1 GHz bandwidth allocation per channel in 60 GHz bands, a SC modulation scheme is a natural choice. Indeed, SC systems have various advantages over their multi-carrier counterparts in implementing 60 GHz systems. First of all, they require less stringent radio design. This is because SC systems are in general less sensitive to power amplifier nonlinearity and RF phase noise as described in Chapter 5 (see Section 5.4). Secondly, SC systems require a smaller effective number of bits (ENOB) in data converter designs. Data converters sometimes can be skipped for systems with non-coherent detection. Thirdly, since the design point of SC systems, in terms of performance and complexity, can be chosen, it can cover a wider range of applications. Given that it is still challenging to implement an orthogonal frequency division multiplexing (OFDM) mode in battery-powered devices, SC modes should be supported by every 60 GHz device, including those with OFDM modes, in order to communicate across different devices. As described in detail in Chapter 6, future technology scaling is naturally expected to improve digital integration and performance but not analog/RF performance in general. For this reason, the relative challenges of implementing 60 GHz OFDM and SC systems with respect to RF non-idealities are not expected to change significantly in the future.

On the other hand, in 60 GHz SC systems, both carrier recovery and symbol timing recovery have to be carefully considered. Even if the carrier phase and frequency recovery can be adequately dealt with in the RF front-end, or can be neglected since non-coherent detection is used, the baseband needs to deal with symbol timing recovery at more than a gigasymbol per second as is shown in Figure 7.1. The sampling points may shift due to clock offset and jitter in the transmitter (TX) and receiver (RX). Analog timing recovery, using clock data recovery (CDR) for example, does not always work, in particular for packet transmission systems. On the other hand, digital timing recovery with extensive oversampling will increase

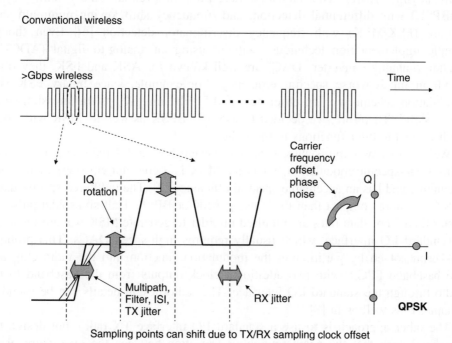

Figure 7.1 Challenges in handling faster than Gbps data rates in wireless systems. Disturbances in output waveform from quadrature demodulator.

the data converter cost and power consumption. In addition, the received signal is shaped by spectral mask requirements as well as radio inherent characteristics, and appropriate equalization schemes to remove the inter-symbol interference (ISI) will become important as the modulation index increases. Furthermore, though it can be reduced with well-controlled beamforming antenna techniques, channel multi-path effects result in additional ISI. Since SC systems need to transmit multi-Gbps data streams without splitting them into multiple sub-carriers, a typical channel delay spread tends to be larger than the symbol period. Still, as long as multi-path effects can be controlled and thus coherent bandwidth is high (i.e. the channel is highly frequency selective) and the nulls in the channel frequency response are not in severe deep fade, SC systems can actually behave quite nicely by spreading frequency-dependent effects across multiple symbols. It is important to note that this argument is often independent of whether the channel is LOS or NLOS.

7.3 System Design with Non-Coherent Detection[2]

In non-coherent detection, the receiver makes no assumption about the carrier phase. In the literature there are three known non-coherent techniques: amplitude shift keying (ASK) with envelope detection, differential binary shift keying (DBPSK) with differential detection, and /frequency shift keying minimum shift keying (FSK/MSK) with frequency discriminator detection [6]. Even though simple implementation techniques without using an analog-to-digital (ADC) or digital-to-analog converter (DAC) are well known for ASK and FSK, they often result in much worse spectral efficiency. For example, since binary FSK is a modulation scheme with an efficiency of 0.5 b/Hz, multi-Gbps transmission could be wasteful. Therefore, we decided to use $\pi/2$ BPSK as well as MSK with some built-in analog filter functions in the radio.

We consider two approaches. One is to design both a TX radio and a baseband with MSK-specific modulation functions. MSK is a constant-envelope modulation technique and has an acceptable spectral efficiency [7]. The unfiltered spectra using MSK are more compact than those using ASK or BPSK. Half-sinusoidal pulses at faster than Gbps data rates are required in order to generate MSK waveforms using a standard I/Q interface, which usually requires high-speed DACs. This problem has been solved by partitioning the modulation function into the radio chip and the baseband FPGA with two additional clock outputs from the baseband to the radio through the standard I/Q interface. The radio desgin details can be found in Chapter 6 as well as in [8].

The other approach is to use a standard I/Q interface TX radio, but design the baseband such that the nearest-neighbor transition between the quadrature phase

---

[2]Reproduced by permission of © 2006, 2008 IEEE.

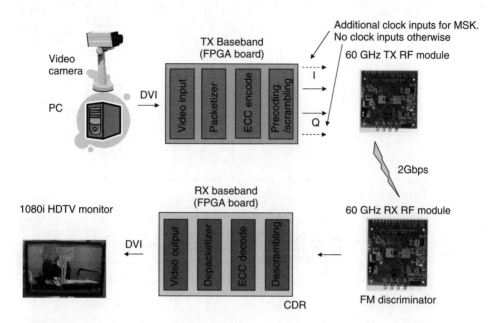

Figure 7.2   High-level block diagram of system with non-coherent detection.

shift keying (QPSK) constellation points can encode MSK-like signals. This approach can be considered as a simple way of generating $\pi/2$ BPSK signals, since, if one is rotated by $\pi/4$ with respect to the other, they are identical. In RX, the I/Q phase is rotated anyway. This approach does not require tristate I/Q signaling or a $\pi/2$ chip-level rotator. It is known that the $\pi/2$ BPSK modulation formats with an appropriate filtering can generate MSK-compatible signals; see [9].

The prototype demonstration system shown in Figure 7.2 is designed to confirm both approaches. It takes HDTV streaming video from a camera or other device in DVI format and transmits it from the SiGe RF TX module to the RX module with appropriate baseband signal processing in each FPGA, and finally outputs the video to an HDTV monitor with DVI input. It functions as a wireless DVI-to-DVI cable. Figure 7.3 shows more detailed block diagrams for the TX/RX baseband. The baseband in each side consists of a single FPGA (Xilinx Vertex-II Pro XC2VP50) as well as the external DVI interface chip and the video buffer memory located on the demonstration board.

Baseband input and output formats are shown in Figure 7.4. In the TX, video sources generate uncompressed HDTV streaming output in DVI format. The streaming image is converted to a 1080i signal (30 Hz frame rate, YCbCr = 4:2:2, 10-bit pixel depth) if necessary, and passed to the packetizer through asynchronous FIFO. The packetizer processes the stream with a 20-bit width at 100 MHz and splits it into a pair of 300-bit packets.

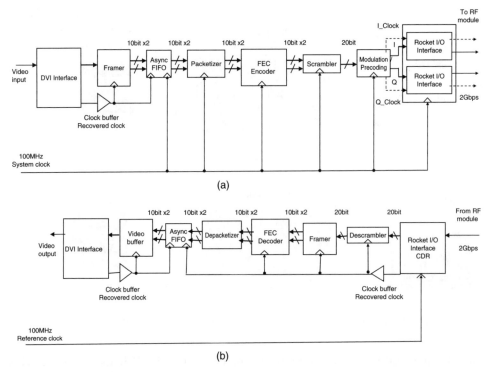

Figure 7.3 Detailed block diagram for non-coherent baseband for DVI video input/output: (a) transmitter baseband; (b) receiver baseband.

Parity symbols are added in the error control code (ECC) encoding block. We designed an experimental codeword configuration that can recover random errors as well as packet losses, since we had some uncertainty in error patterns caused by both impairments and disturbances in the RF section and the channel. The scheme is used for both non-coherent and differentially coherent systems. For random error correction we use a BCH (350, 300) code that can correct up to five random bit errors, and for burst errors and packet losses we use symbol-based partially over-lapped block (POB) code (POB (150; 149; 3)) [10]. Both codes are built on GF(2^{10}) and combined as a concatenated code. The data is transmitted, after appropriate pre-coding and scrambling (with $x^9 + x^4 + 1$), to the TX RF module through FPGA's high-speed I/O ports. We avoid using convolutional codes, since the code rate is low and burst error correction capability is limited unless deeply interleaved. Our block-code-based approach enables good error correction performance for both random and burst/packet errors at high code rate. The detailed configuration for BCH (350, 300) and POB (150; 149; 3) is shown in Figure 7.5. For the present POB code, an $n = 150$ symbol frame is divided into $s = 3$ subframes (each containing

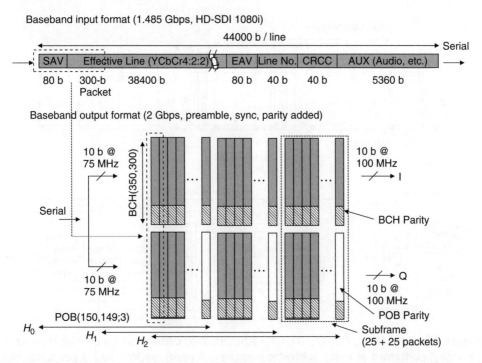

Figure 7.4   TX baseband input/output format. Input DVI stream shown above is framed into two streams of data, and parity data for both inner and outer codes are added.

$n/s = 50$ symbols: 25 data packets for I and 24 data $+$ 1 parity packets for Q, respectively) and a set of three (150, 149) Reed–Solomon (RS) codes (the parity check conditions are generated by 1, α, and α^2, respectively) are used to recover lost packets.

The RX baseband also consists of a single FPGA as shown in Figure 7.3. No ADC board is used and the digital signal of the FM discriminator output is transferred to a CDR of the FPGA. After the clock and data are separated, the signal remapping and descrambling are done. The BCH inner decoder corrects random errors and detects corrupted packets and passes the signal to the outer decoder. The outer decoder recovers packets based on the uncorrectable packet signals from the inner decoder. After random error correction and packet recovery, packets are reformatted as video lines after the parity symbols are stripped off in the depacketizer. The signal is passed through asynchronous FIFO to adjust the clock frequency differences, similar to the transmitter baseband. If the cyclic redundancy checksum does not match, that particular line is discarded, and the video retraces the last correctly received line. The video output section converts the video stream to DVI format and outputs it to the HDTV monitor.

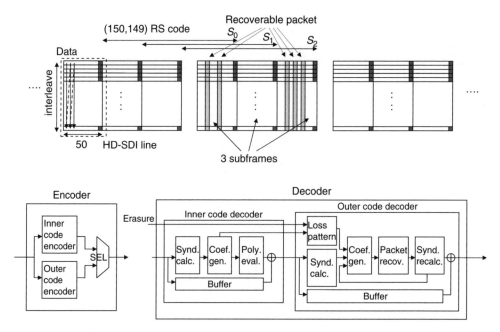

Figure 7.5 (a) POB(150, 149; 3) uses 3 RS(150, 149) constituent codes with $(1, \alpha, \alpha^2)$. Decoding performed at every subframe boundary, 3 parity packets added per frame but can recover up to 5 packets per frame. (b) Decoder block diagram, showing inner code decoder for random error correction and outer code decoder for packet recovery.

A block diagram of the decoder is shown in Figure 7.5. The BCH inner decoder corrects random errors, detects corrupted packets and passes the signal to the outer decoder. The BCH decoder circuit consists of units for syndrome calculation, error locator polynomial generation, and Chien search. The outer decoder recovers packets depending on the uncorrectable packet signals from the inner decoder.

Although the non-coherent detection system can make use of CDR to recover bit timing for streaming applications for low-cost and low-power implementations in relatively clean channel environments, the maximum data rate for a given bandwidth is often reduced by half. This is because we cannot separate the information transmitted in the in-phase (I) and (quadrature) Q channels effectively. In addition, the lack of data converters does not allow us to consider potential extensions for more advanced and robust baseband signal processing techniques for non-negligible multi-path effects, etc. Furthermore, even though per-packet carrier phase recovery is not required, per-packet symbol timing recovery will be needed for packet transmission systems, and the symbol timing acquisition will need to be improved with digital techniques.

7.4 System Design with Differentially Coherent Detection[3]

With these aspects in mind, an alternative system implementation can be considered. Figure 7.6 shows the block digram of our differentially coherent demonstration system. DQPSK modulation is a 2-bit/symbol phase modulation scheme mapping differentially encoded binary information in the two packet streams into the I and Q channels. These signals are delivered to the TX RF module at a data rate of 1 Gbps. Even though our radio can support other modulation methods such as ASK and BPSK, it is difficult to pass the data in I and Q separately in those cases.

Similar to our non-coherent demonstration, the system takes an HDTV image from a camera and transmits it through SiGe RF modules with appropriate baseband signal processing, and outputs it to an HDTV monitor.Although the use of a digital sampling approach with high-speed ADC will allow us to adopt more advanced signal processing techniques, the receiver-side timing recovery will become more difficult, causing more frequent sync/burst errors. We therefore added preamble and distributed synchronization patterns to help recover timing information as shown in Figure 7.7, when the data is passed to the DQPSK modulation block. Exploiting the FEC design described in the previous section can help in these aspects. In other respects the transmitter design is straightforward DQPSK baseband.

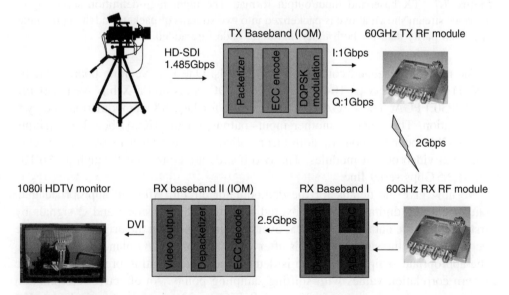

Figure 7.6 DQPSK demonstration system overview. A 2 Gbps data stream is transmitted over the 60 GHz SiGe radio link.

---

[3]Reproduced by permission of © 2007 IEEE.

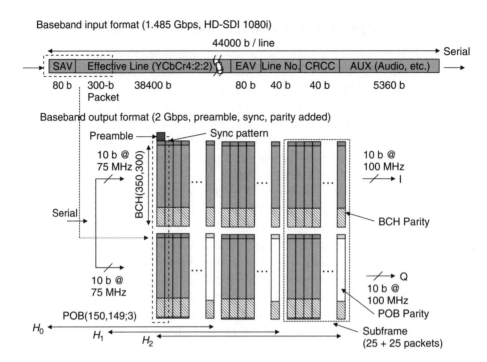

Figure 7.7 TX baseband input/output format. The input high-definition serial digital interface stream shown above is packetized into two streams of packets and then preamble, sync, and parity data for both inner and outer codes are added.

The receiver baseband consists of two boards. One is a Neptune board (Qinetiq VXS1) with dual 2 Gsps 10-bit high-speed ADC chips and a Xilinx Vertex-II Pro XC2VP50 FPGA. The FPGA contains circuits for DQPSK demodulation and synchronization. The other is another input–output module (IOM) board, containing the ECC decoder for both random error correction and packet recovery, depacketizer, and video output modules. The two boards are connected through an 8b/10b coded 2.5 Gbps serial line.

We now give the details for the demodulator circuit. After an appropriate gain adjustment, the digital data is recovered from the receiver-side I and Q signals by first sampling the data at 2 Gbps (2× oversampling) and then the original data can be recovered from the sampled signals after adjusting for the I/Q plane phase rotation. The appropriate sampling position is determined by calculating preamble and sync pattern correlation values with shifting sampling points. An effective interpolation scheme is used, as is shown in Figure 7.8. The correlation values are calculated at the two oversampling points (A and C). If these correction values are too close, we can tell that it is more appropriate for the sampling point to be close to B. Therefore, the output data was generated by interpolating the data at A and C. This

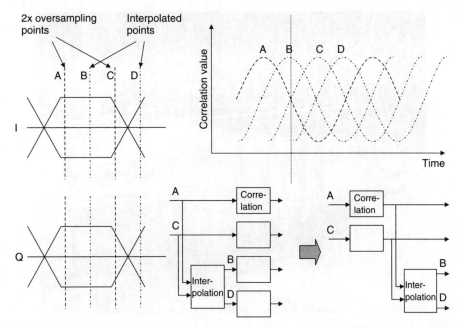

Figure 7.8 Symbol-timing recovery technique with simple but effective interpolation in the demodulator block. The use of interpolated points, B and D, in addition to sampled points A and C, can give more stable results. Performing the interpolation after the correlation value calculation for synchronization can reduce the circuit size and power consumption. I and Q signal mixture is ignored for simplicity.

scheme avoids calculating correlation values at four different points (two sampled, two interpolated), so the circuit size and power can be reduced.

Then the demodulation output is passed to another baseband board (output IOM board), where we perform error correction to recover for random errors and packet losses.

7.5 Test and Evaluation[4]

Figure 7.9 shows a setup of our demonstration system for uncompressed video streaming. In order to achieve a successful demonstration efficiently, we split the demo construction and testing into three steps: first the baseband designs were tested with wired connections; then the radio modules were added for wireless testing, and finally, end-to-end testing and measurements were performed with video I/O.

The baseband testing was performed by connecting the output of the transmitter I and Q signals into the input of the receiver I and Q after passing through appropriate

[4]Reproduced by permission of © 2007, 2008 IEEE.

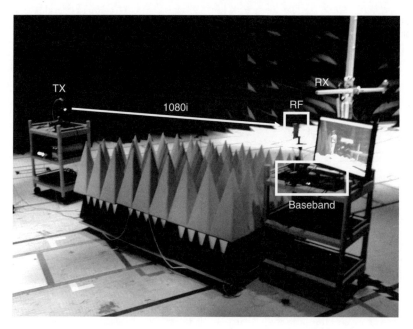

Figure 7.9 Wireless transmission demonstration system in a radio anechoic chamber (camera, attached in TX, not shown).

attenuators for the differentially coherent system. For the non-coherent system, the transmitter-side interface was changed for wired testing. Since the present demonstration systems are built in such a way that the RF and baseband are physically separated, they are sensitive to the cable length of the I and Q channels. They need to match perfectly, or otherwise the signal constellation becomes severely affected. This step also helped us to find various bugs in the digital baseband design, such as packet framing and coding, prior to connecting to the RF modules. It is often difficult to separate hidden logic bugs from inherent bit errors in wireless environments, and wired transmission under controlled channel environment is mandatory to speed up the debugging phase without running logic simulations for a long period of time.

After confirming the baseband operations in wired configuration, the RF modules were added, and testing with the RF modules was performed with some testing patterns. Again, the cable lengths were carefully adjusted. We verified the signal transmission by monitoring the signals in the FPGA with an appropriate software tool. The data was collected by using the data capture capability of the ADC board.

Finally, the source was switched to the camera, and end-to-end system verification was performed. Throughout our demonstration design experience, we found that the FPGA platform really helps to adjust the design to compensate for unavoidable disturbances in the RF front-end and channels.

Random error dominant (Phase II)

Figure 7.10 Wireless transmission results for error pattern measurements in the non-coherent system. Note that measurement is performed by disabling FEC.

We found that the present video streaming demonstration systems are attractive not only for "proof-of-concept" demonstration but also as a powerful measurement tool. This is mainly because error patterns are visible in real time. For example, for a non-coherent system under an multi-path suppressed environment, random errors (white dots) are dominant as shown in Figure 7.10. Random error correction is quite effective in such a case. Although a non-coherent system with no data converter implementation excludes enhanced options such as RAKE and equalization in the digital domain, it can still be the basis of a practical system, in particular when the multi-path effect is small (e.g. the distance is short or the antenna gain is large).

Figure 7.11 shows how video robustness is affected for demodulator and FEC design for the differentially coherent system. When both ECC and interpolated sync are turned off, the image quality is reduced by many burst errors which we think mainly arise from synchronization errors (Figure 7.11(a)). When either interpolated sync or error correction is added (Figure 7.11 (b) or (c)), the burst errors are reduced, though not completely. When both functions are turned on (Figure 7.11(d)), clear and crisp images are obtained. Since random errors are not very noticeable, the coding overhead in the present design is expected to be reduced by using a simpler BCH code. We also note that we were successful in some NLOS transmissions by blocking the LOS path and using a specular reflection signal instead.

7.6 Advanced SC System with Per-Packet Coherent Detection

In this section we will show how to implement an advanced SC system that is compliant with a given standard such as IEEE 802.15.3c. Though detailed discussion of various design options is beyond the scope of this chapter, there are several important considerations.

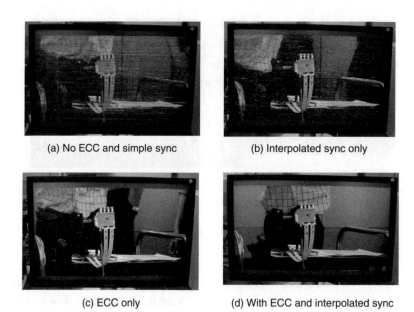

(a) No ECC and simple sync   (b) Interpolated sync only

(c) ECC only   (d) With ECC and interpolated sync

Figure 7.11   Wireless transmission results for differentially coherent system: (a) no ECC and simple sync; (b) interpolated sync only; (c) ECC only; (d) both ECC and interpolated sync turned on.

First of all, the coherence needs to be maintained between transmitter and receiver on a per-packet basis, since no idle patterns can be sent between packets in multiple-access configurations. As a result, carrier phase and frequency recovery as well as symbol timing recovery need to be performed on a per-packet basis. Analog carrier recovery techniques such as the Costas loop [11] are well known, but frequency acquisition takes time. For faster acquisition, digital-domain techniques are more preferable [12]. But it is still a challenge to maintain the phase coherence per packet in 60 GHz, since everything is an order of magnitude faster. It should be noted that the signal processing latency can be a major limitation in system loop bandwidth. The preamble detection circuitry has to quickly extract necessary information relating to carrier phase and frequency offset, symbol sampling phase offset, channel condition, etc. Efficient implementations are known in Golay correlators [13], and the design may need additional enhancements for higher data rates.

Secondly, equalizer design is critical in designing robust systems. The use of frequency-domain equalization has been discussed in detail in Chapter 5 as well as in [14], but the complexity will become comparable to or even exceed that of OFDM. The lack of pilot tones in SC systems may result in additional difficulties in singnal recovery. Alternatively, time-domain equalizers are an interesting option with high-gain antennas, considering the relationship between delay spread and antenna gain as

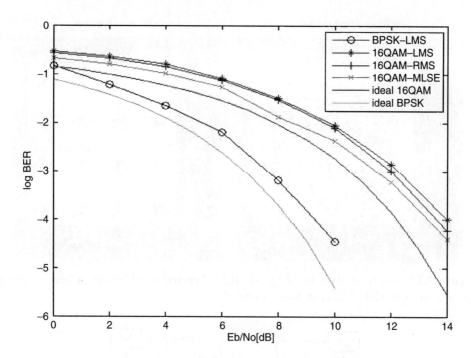

Figure 7.12   Equalizer design comparison.

noted in Chapter 2. In other words, as antenna gain becomes higher, the delay spread is expected to be reduced since fewer multi-path components are involved. If the delay spread is shorter, the number of time-domain equalizer taps can be reduced. As a result, the time-domain equalizer can be a more efficient choice than the frequency-domain equalizer.

When implementing a time-domain equalization, we need to further consider which time-domain equalization method to chose, and we need to carefully compare the performance – complexity tradeoffs in different methods. Figure 7.12 compares various time-domain equalizer implementations using MATLAB [15]. A static-filter-only solution is not appropriate in real applications, since it is known to be sensitive to sample timing offset. On the other hand, maximum smoothed likelihood estimator filters result in high complexity. We consider that adaptive filters such as least mean-square filters are appropriate. The advantage for 16-QAM is shown in Figure 7.13 where an adaptive digital filter is designed to recover a SC 16-QAM constellation [15].

Thirdly, decoding latency requires more attention in FEC designs, since, for example, headers are protected with RS(33, 17) code and the decoding needs to be finished before starting the signal processing in the payload part. Low-latency

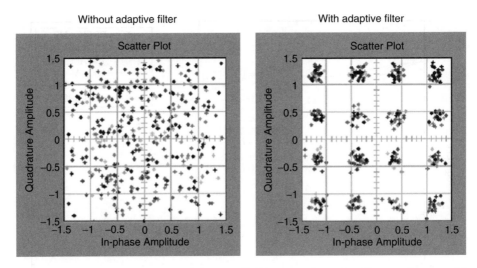

Figure 7.13 Improvements in 16-QAM. IEEE spectral mask effect, a few ns delay spread, additive white Gaussian noise included.

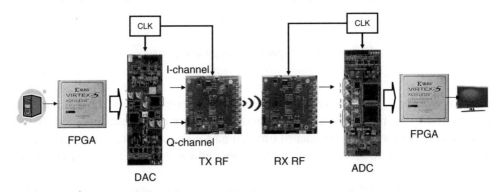

Figure 7.14 FPGA platform configuration for advanced SC systems.

FEC design techniques [16] can be of help. Low-density parity check decoders also require high performance and low latency design. Lower latency circuit design and implementation will also critical to improve the system tracking by increasing loop bandwidth.

Finally, for advanced SC system demonstration, an FPGA platform equipped with high-speed DAC/ADC such as shown in Figure 7.14 will play an important role. Although digital designs can be verified by means of simulation, analog impairments and disturbances in radio, antenna, channels are hard to model exactly and measurement feedbacks are critical in finalizing various design options. Also, it can help to flexibly adopt moving standards.

Table 7.1 Comparison for various 60 GHz SC implementations described in this chapter

	Non-coherent	Differentially coherent	Coherent
Modulation	$\pi/2$-BPSK, MSK	DQPSK	$\pi/2$-BPSK, $\pi/2$-QPSK
PHY frame	Proprietary	Proprietary	IEEE 802.15.3c
Data converters	Optional	Needed	Needed
Roadmap	None	None	$\pi/2$-8PSK, $\pi/2$-16QAM

7.7 Conclusion

We have discussed SC system implementation examples, starting with two proof-of-concept implementations, one with non-coherent detection and the other with differentially coherent detection, and then discussing more advanced system design that achieved per-packet coherent detection. A comparison summary is given in Table 7.1.

Since one of the biggest advantages of SC schemes is wide application coverage for both AC-powered and battery-operated devices, it is quite important to show that the scheme constitutes a roadmap for higher data rates. Our approach with advanced SC baseband design looks promising because it can manage per-packet coherence for higher data rate modulation schemes such as 16-QAM.

As discussed, the advantage of the SC modulation scheme will be augmented with phased array antenna techniques, since multi-path can be actively controlled to focus on the strongest beam path available, whether it is LOS or NLOS, to reduce the channel delay spread.

References

[1] Katayama, Y., Haymes, C., Nakano, D., Beukema, T., Floyd, B., Reynolds, S., Pfeiffer, U., Gaucher, B. and Schleupen, K. (2007) 2-Gbps uncompressed HDTV transmission over 60-GHz SiGe radio link. *IEEE Consumer Communications and Networking Conference (CCNC)*, January.

[2] Katayama, Y., Nakano, D., Valdes-Garcia, A., Beukema, T. and Reynolds, S. (2008) Multi-Gbps wireless systems over 60-GHz SiGe radio link with BW-efficient noncoherent detections. *IEEE International Conference on Multimedia and Expo* (ICME), June.

[3] Katayama, Y. (2009) End-to-end 60 GHz single-carrier system implementation and link experiments, *IMS Workshop on System-Level Design and Implementation of Gb/s 60 GHz Radios*, June.

[4] Nakano, D., Kohda, K., Yamane, T., Ohba, N. and Katayama, Y. (2010) Robust 60 GHz single-carrier system for per-packet coherent detection. Preprint.

[5] Ohba, N., Kohda, Y., Nakano, D., Takano, K., Yamane, T. and Katayama, Y. (2009) *FIT 2009*, September (in Japanese).

[6] Rappaport, T.S. (2002) *Wireless Communications: Principles and Practice*, 2nd edn. Upper Saddle River, NJ: Prentice Hall PTR.

[7] Pasupathy, S. (1979) Minimum shift keying: a spectrally efficient modulation. *IEEE Communications Magazine*, **17**(4), 14–22.

[8] Reynolds, S., Valdes-Garcia, A., Floyd, B., Gaucher, B., Liu, D. and Hivik, N. (2007) Second generation transceiver chipset supporting multiple modulations at Gb/s data rates. *IEEE Bipolar/BiCMOS Circuits and Technology Meeting*, pp. 192–197, October.

[9] Lakkis, I., Su, J. and Kato, S. (2001) A simple coherent GMSK modulator. *IEEE International Symposium on Personal, Indoor and Mobile Radio Communications (PIMRC)*, pp. A-112–A-114, September.

[10] Katayama, Y. and Nakano, D. (2006) Multiple-packet recovery technique using partially-overlapped block codes. *International Symposium on Information Theory* (ISIT), July.

[11] Costas, J.P. (1959) Synchronous communications *proceedings of the IRE*, **47**, 2058–2068.

[12] Sari, H. and Moridi, S. (1988) New phase and frequency detectors for carrier recovery in PSK and QAM systems. *IEEE Transactions on Communications*, **36**(9).

[13] Popovic, B.M. (1970) Efficient Golay correlator. *Electronics Letters*, **35**(17).

[14] Sari, H., Karam, G. and Jeanclaud, I. (1994) Frequency-domain equalization of mobile radio and terrestrial broadcast channels. *Global Telecommunication Conference*, vol. 1, pp. 1–5.

[15] Yamane, T. (2009) Personal communication.

[16] Katayama, Y. and Yamane, T. (2003) Concatenation of interleaved binary/non-binary block codes for improved forward error correction. *Optical Fiber Communications Conference* (OFC), March.

8

Gbps OFDM Baseband Design and Implementation for 60 GHz Wireless LAN Applications

Chang-Soon Choi, Maxim Piz, Marcus Ehrig and Eckhard Grass

For high data-rate wireless transmission over non-line-of-sight (NLOS) 60 GHz channels, the orthogonal frequency division multiplexing (OFDM) scheme has been preferred to single carrier (SC) schemes due to its stronger immunity to highly frequency-selective fading [1].[1] This would make baseband receiver implementation easier since costly equalizers required in SC schemes can be significantly simplified. However, an OFDM is more sensitive to synchronization errors compared to a SC scheme. As an additional challenge, 60 GHz phase-locked loops (PLLs) exhibit worse frequency instability and phase noise characteristics than microwave-band ones. Therefore, it is necessary to design synchronization architecture more carefully in a 60 GHz OFDM baseband receiver.

This chapter describes design consideration and implementation issues for the 60 GHz OFDM baseband processor which we developed for the WIGWAM [2] and EASY-A projects [3]. An FPGA platform has been utilized for baseband implementation since it allows simpler modification as well as higher flexibility/scalability in system-level hardware. Moreover, it promises simple transitions to structured application-specific integrated circuit (ASIC) chips.

[1]This work was funded by the German Federal Ministry of Education and Research (BMBF) through the WIGWAM and EASY-A projects. The authors would like to thank the analog circuit team at IHP.

---

60 GHz Technology for Gbps WLAN and WPAN: From Theory to Practice
Edited by Su-Khiong (SK) Yong, Pengfei Xia and Alberto Valdes Garcia
© 2011 John Wiley & Sons, Ltd

However, an FPGA implementation for multi-gigabit data rate is not a straight-forward task due to limited clock speed. The most critical blocks are fast Fourier transform (FFT) and Viterbi decoding which require a high clock rate not avail-able in FPGA. For example, the first demonstrator targeted more than 1 Gbps data throughput and this would lead to Viterbi decoder hardware operating at 1 GHz. However, this is not available in any FPGA platform and even in ASICs with a stan-dard design approach. Therefore, a highly parallelized architecture is indispensable for higher data throughput than Gbps with FPGA platform.

This chapter is organized as follows. Section 8.1 describes the designed OFDM physical layer and frame architecture. In Sections 8.2 and 8.3 the baseband processor architectures and their implementation details for OFDM receivers and transmitters are presented. Section 8.4 is dedicated to an indoor wireless link demonstration with the developed 60 GHz radio front-ends. Finally, the next-generation OFDM demonstrator and its performance evaluation are presented.

8.1 OFDM Physical Layer Implemented on FPGA

8.1.1 Designed OFDM Physical Layer

The key OFDM timing parameters used in the baseband developed are summa-rized in Table 8.1. We assumed a channel bandwidth of 500 MHz for one channel, therefore parameters were adapted to a bandwidth of 330 MHz with the sampling frequency of 400 MHz. Cyclic prefix length and sub-carrier spacing were chosen to yield a good compromise between tolerated delay spread in 60 GHz channels and phase noise immunity to a 60 GHz PLL fabricated on IHP SiGe technology [5].

Table 8.1 OFDM timing parameters for the 60 GHz demonstrator developed [4]. Reproduced by permission of © 2010 IEEE

Parameter	Value
Channel bandwidth	500 MHz
FFT bandwidth	400 MHz
FFT size	256
Sub-carrier spacing	1.5625 MHz
Guard interval	160 ns ($N_G = 64$)
FFT period	640 ns
OFDM symbol time	800 ns
Data sub-carriers	192
Pilot/Zero sub-carriers	16/5
Nominal used bandwidth	333 MHz

There have been several reports that 60 GHz wireless channels exhibit root-mean-squared (rms) delay spreads ranging from a few nanoseconds to several tens of nanoseconds in a typical indoor environment [6–9]. In general, an OFDM system is designed to have a cyclic prefix length greater than twice the rms delay spread in a wireless channel. Moreover, filters would add up channel responses, causing the effective guard interval to be reduced. Consequently, a guard time of more than 150 ns would be required to support 64-QAM modulation. To avoid substantial loss of data rate, the guard time should be only a small fraction of the OFDM symbol time (e.g. 20% in IEEE 802.11a). To lower the impact of the guard time on efficiency, we could have increased the symbol duration using a larger FFT. However, this approach would lead to higher susceptibility to phase noise because of smaller sub-carrier spacing. As a result, a cyclic prefix of 160 ns together with a FFT period of 640 ns (256 sub-carriers in 400 MHz bandwidth) resulting in a symbol time of 800 ns were chosen. Unused guard sub-carriers near the channel edge facilitate channel filtering with low complexity. The convolutional coding (171, 133) was utilized for forward error correction. With defined puncturing patterns and modulation from binary shift keying (BPSK) to 64-QAM, eight data modes with different data rates are supported as shown in Table 8.2. Bit interleaving was performed over data blocks equal to one OFDM symbol.

To increase immunity to 60 GHz channel and phase noise impairments, we developed a novel synchronization and channel estimation scheme. Detailed information on this scheme, including a performance evaluation, is available in [10]. With the preamble structure shown in Figure 8.1, it allows sufficiently precise time and frequency synchronization performance without a costly cross-correlator. The preamble has a length of 11 OFDM symbols and it enables robust initial synchronization and channel estimation despite high phase noise arising from 60 GHz PLLs. To achieve high efficiency, the data frame length should be chosen sufficiently large using medium access control (MAC) frame aggregation.

Table 8.2 Modulation and coding scheme

Modulation	Code rate	Data rate (Mbps)
BPSK	1/2	120
BPSK	3/4	180
QPSK	1/2	240
QPSK	3/4	360
16-QAM	1/2	480
16-QAM	3/4	720
64-QAM	2/3	960
64-QAM	3/4	1080

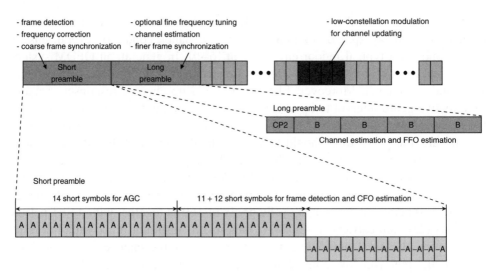

Figure 8.1   Preamble architecture used in OFDM baseband implementation.

8.1.2 Performance Evaluation in the Presence of Clock Deviation and Phase Noise

Figure 8.2 shows the simulated performance for 16-QAM modulation with 1/2 convolutional coding rate, providing 480 Mbps data rate. A frame error rate (FER) curve is plotted for a frame length of 2048 bytes. We accounted for all overhead losses including preamble, signal field, pilot signals and cyclic prefix. The channel model used in this simulation was developed for 60 GHz LOS channels realized with an omnidirectional antenna at a transmitter and a Vivaldi antenna having 30° half-power beamwidth (HPBW) at a receiver. This performance is compared with an ideal case where no phase noise and sampling timing drift is applied [11]. For the practical case scenario, we have assumed 30 ppm clock frequency deviation and a Wiener model for the phase noise. Both transmitter and receiver PLLs were assumed to have a single sideband phase noise value of $L_{SSB} = -93$ dbc/Hz at 1 MHz offset. It is observed that a frame error rate of 10% is achieved at an energy per bit to noise power spectral density ratio (E_b/N_0) of 12.3 dB. The performance loss due to phase noise and sampling timing drift is about 0.7 dB.

8.2   OFDM Baseband Receiver Architecture

An OFDM baseband processor was implemented with FPGA platform using Virtex-II pro devices [11]. Prior to hardware implementation, as exemplified in Figure 8.2, we performed extensive simulations of the designed OFDM baseband under different channels and hardware impairments. Afterward, the complete transceiver was

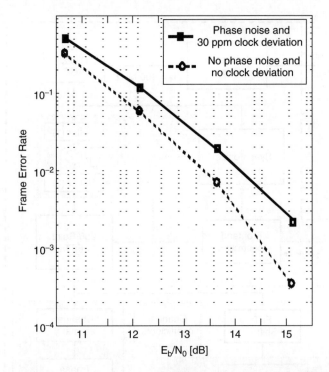

Figure 8.2 Performance evaluation for 480 Mbps data mode (16-QAM, 1/2 convolutional coding) [11]. Reproduced by © 2009 IEEE.

synthesized and mapped to FPGA. The synthesis results indicated that the complexity of the OFDM receiver architecture is more than twice higher than that of the transmitter due to the high complexity of the Viterbi decoder as well as the additional blocks required for receiver operation. Therefore, we describe the implementation of an OFDM receiver first in this section.

In Figure 8.3, a block diagram for the developed OFDM receiver is shown. The OFDM receiver can be functionally divided into an inner receiver (or digital front-end) and an outer receiver (or data-path processor). The main function of a digital front-end is to process the incoming I/Q-baseband signal waveform and to retrieve constellation symbols which are mapped on sub-carriers. This digital front-end also includes the receiver functions required for OFDM signal processing, such as synchronization, FFT operation, channel estimation, channel equalization, common phase correction and phase tracking. The data-path processor performs symbol demapping/weighting, deinterleaving and channel decoding. As mentioned earlier, there was the hardware limitation that the typical clock speed achievable for complex designs mapped on the Virtex-II pro family was around 100 MHz. In order to support a system bandwidth with 400 Msps, four signal samples or

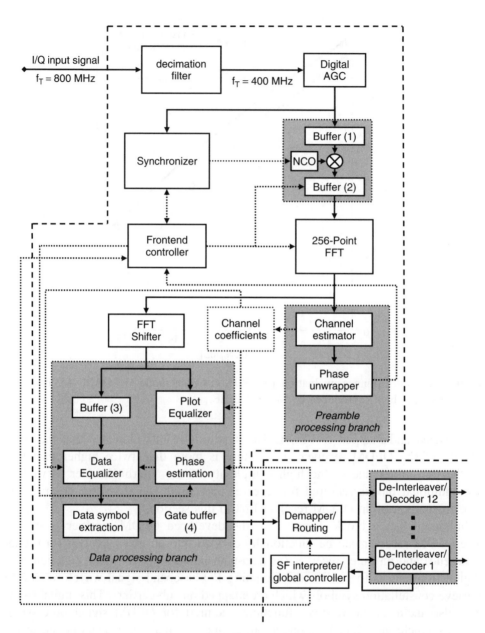

Figure 8.3 OFDM receiver architecture [11]. Reproduced by © 2009 IEEE.

sub-carrier symbols were multiplexed and processed in parallel in most of the digital front-end stage. Full pipelining was also employed to achieve the required total throughput. The processing in the blocks after and including the FFT process is burst-oriented. To process gigabit throughput with available Viterbi decoders, the OFDM transmitter de-multiplexes the source stream into 12 sub-streams, which are encoded separately. The receiver decoding block contains a bank of 12 parallel decoding blocks. Each block consists of a deinterleaver and a Viterbi decoder with maximum processing up to 100 Mbps. The number of blocks involved during a frame reception depends on the data mode.

8.2.1   Receiver Front-End

8.2.1.1   Input Stage and Synchronization

The complex-valued I/Q signals from the 60 GHz analog front-end are first oversampled with an 8-bit analog-to-digital (ADC) converter at a sampling rate of 800 Msps. A finite impulse response half-band filter then performs decimation by a factor of 2. A digital automatic gain control (AGC) amplifier ensures a defined input level for subsequent stages. The designed synchronizer performs frame detection, coarse timing estimation and carrier frequency offset (CFO) estimation with the defined preamble architecture [10]. The AGC gain level is kept fixed after frame detection to prevent further amplitude variations. Due to processing latency, frequency offset information is only available after the second preamble part has already started. Therefore, a buffer stage is included in parallel to the synchronizer. As soon as the frequency of a numerical controlled oscillator (NCO) has been set, the system starts to continuously read out the first FIFO buffer, corrects a frequency shift by complex multiplication with a rotating phasor and writes the result in a second FIFO buffer. When setting the frequency offset, the synchronizer also announces a new successfully detected frame and provides the start index for the first FFT.

8.2.1.2   Fast Fourier Transform

Due to the limited clock speed, the radix-4 algorithm was used to realize a four-port 256-point FFT block with four 64-point FFTs working in parallel; see the block diagram in Figure 8.4. This enables us to process samples with 400 Msps with a 100 MHz FFT processor at the expense of area. In accordance with the algorithm, the output results of these sub-FFTs are fed into a multiplier stage, followed by a radix-4 butterfly and a reordering unit. The multiplier stage performs multiplication with the complex twiddle factors. The pipelined architecture offers maximum throughput without any waiting cycles.

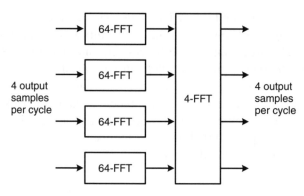

Figure 8.4   FFT block diagram.

8.2.1.3   Front-End Controller

The front-end controller manages to control global flows in the digital front-end. It features a state machine for sequencing and initiation of FFT operations at defined times. Transfers of data bursts from the input buffer 2 into the FFT are managed by a sub-controller. Such a data burst read can be initiated as soon as the first sample is available in buffer 2. The front-end controller selects whether FFT output data is delivered to the preamble processing or the data processing branch.

8.2.1.4   Scheduling

The front-end controller needs to wait until the fine timing position is estimated by the phase unwrapper. With readjusted FFT positioning, the receiver starts the processing of accumulated waveform data. The first data symbol in the frame constitutes the signal field which defines transmission mode parameters. They become available when the signal field has passed through the complete receiver chains including the data processing branch in the front-end and the demapper, deinterleaver, decoder and signal field interpreter. Given a transmission without immediate acknowledgment, some inter-frame guard time is necessary for a receiver to be able to process the next transmitted frame. Two provisions keep this guard time short. Firstly, instead of waiting for a reply from the interpreter, the receiver immediately starts to process the subsequent accumulated data symbols regardless of the actual frame length. A buffer stage right before the front-end output acts as a gate for data flow. It lets the signal field go through but keeps the next OFDM symbols in the queue until frame length information is available. For short frames or in the event of an erroneous signal field, some stored symbols will be discarded. Secondly, the receiver is able to handle two consecutive frames simultaneously. As soon as the last OFDM symbol is sent to the data processing branch, the front-end controller reactivates the AGC and synchronizer. While the rest of the receiver may still process the older frame, synchronization and channel estimation can already be performed for the next frame. Every OFDM symbol carries a frame label, which is used to allocate the dedicated channel data.

8.2.1.5   Channel Estimation and Fine Timing Synchronization

After the synchronization process, four long preamble symbols are transformed into the frequency domain by FFT and delivered to the channel estimator. These four symbols promise 6 dB channel estimation gain. In general, OFDM provides an easier way to cope with multi-path propagation than SC schemes. Provided that guard time is longer than the channel impulse response and time synchronization is properly established, channel convolution in the time domain transforms into complex multiplication in the frequency domain. Each sub-carrier symbol appears multiplied with the channel transfer function taken at the frequency of the sub-carrier. Like IEEE 802.11a, the long preamble symbols are generated from some pseudo-random BPSK reference sequence $b(k) \in \{-1, 1\}$ in the frequency domain. To estimate the channel coefficients, the receiver flips the sign of those sub-carrier symbols where $b(k) = -1$. Since the receiver employs a simple feed-forward scheme without channel re-estimation, it is required to have the estimated channel coefficient $\hat{H}_i$ and not channel coefficients themselves. The receiver calculates polar representations $\hat{A}_i \exp(j\hat{\phi}_i) = \hat{H}_i$ of the coefficients and stores the inverse channel coefficients as $(1/\hat{A}_i) \exp(-j\hat{\phi}_i) = 1/\hat{H}_i$ at this point. For this purpose, a pipelined four-port Coordinate Rotation Digital Computer (CORDIC) stage and a divider stage were used. In addition, sub-carrier power gain values, estimated as $\hat{P}_i \approx \hat{A}_i^2$, required in the following demapper/weighting and phase estimation blocks.

Fine frame synchronization is accomplished with the subsequent phase unwrapper. Using the time shifting property of the FFT [11], the current FFT window location with respect to the maximum power point is proportional to the average phase increase in frequency domain, which can be estimated by

$$\Delta n \approx \left[\sum_{i \in M} (\hat{\phi}_{i+1} - \hat{\phi}_i) \right] \alpha/\pi, \qquad (8.1)$$

where $\alpha = N_{FFT}/2N_p$ is a constant. The summation is carried out over the set M of those used data sub-carriers ($N_p = 206$), with index i denoting sub-carrier number. The OFDM receiver corrects frame timing by choosing a fixed offset in advance of the channel power maximum. This results in some sample offset Δm from the new to the old FFT window position. Due to the shifting property, this means that we require not the inverse of the original channel estimates $\hat{H}_i$, but the inverse of corrected channel estimates $\hat{H}_i \exp(2\pi j \Delta m \cdot i/N_{FFT})$, for channel equalization.

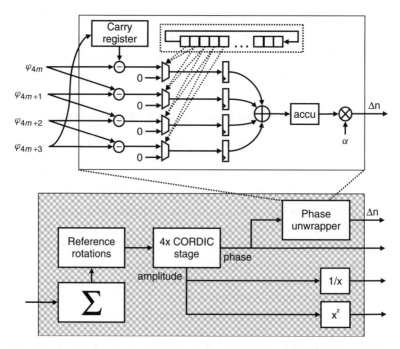

Figure 8.5   Preamble processing branch [11]. Reproduced by © 2009 IEEE.

Having the channel coefficients stored in polar representation, the transformation to the new set of channel coefficients is trivial and easily done in the equalizer. The receiver only needs to store the time offset as Δm a global parameter.

Hardware implementation was realized as shown in Figure 8.5. A 1-bit circular shift register of length 256 is used to identify used sub-carriers. This approach offers low complexity with four multiplexers, some subtractors/adders, registers and some final multiplication. Typically, a cross-correlator used for fine synchronization with a pattern length of N requires N additions per input sample. On the other hand, this scheme requires only one subtraction and two additions per input sample.

8.2.1.6   Phase Correction and Equalization

The OFDM demonstrator presented here was designed for a static 60 GHz channel environment implying a time-invariant channel impulse response during frame transmission. On the other hand, the rapidly changing carrier phase due to PLL phase noise and residual carrier frequency offset induces inter-carrier interference (ICI) and a common phase error (CPE) for all sub-carriers for the nth OFDM symbol. In addition, the symbol clock deviation between transmitter and receiver causes a drift $\Delta\tau(n)$ of the FFT window position with respect to the transmitted symbols.

Due to the FFT time shifting property the sub-carrier dependent phase error $\theta(n, k)$ for OFDM symbol n and sub-carrier $k, k = -128, \ldots, 127$ (offset representation), is equal to

$$\theta(n, k) = CPE(n) + \Delta\tau(n)(2\pi/(T \cdot N_{FFT})) \cdot k \tag{8.2}$$

where $T = 1/f_T$ and f_T denotes the sampling rate. Therefore, estimation of phase error for some OFDM symbol (n) can be done with a line fitting procedure in the phase domain, where the phase error CPE (n) at DC and the phase slope $\Delta\theta(n) = \Delta\tau(n)(2\pi/(T \cdot N_{FFT}))$ are estimated. We make use of transmitted pilot sub-carriers modulated with BPSK for this purpose. These pilot sub-carriers lie on a regular grid with distance $\Delta_p = 14$ between adjacent pilot sub-carriers. To begin with, the pilots are corrected in amplitude and phase by a dedicated pilot equalizer. The BPSK symbols are also demodulated. Pilot equalizer and consecutive pilot processing blocks are reduced in complexity, because only one pilot needs to be processed within a cycle at most. For estimation of the phase parameters, we incorporate reliability information in form of pilot power levels to account for the different sub-carrier signal-to-noise ratios (SNRs) due to the channel. The equalized and unmodulated (sign-flipped) pilot symbols for OFDM symbol n are denoted by $Z_i(n), i = 1, \ldots, 16$, the pilot power values P_i by. Following a heuristic scheme, which can be implemented with low complexity, the phase slope can be estimated by the equation

$$\Delta\theta(n) = \left(\angle \left[\sum_{i=1,\ldots,15} \bar{Z}_i(n) Z_{i+1}(n) \cdot \min(P_i, P_{i+1}) \right] \right) / \Delta_p. \tag{8.3}$$

This equation can be regarded as a phase unwrapping scheme with power weighting. Simulation showed that link performance was very sensitive to estimation precision of the phase slope and the variance of the estimator was too high. As a result, link quality would strongly degrade from inner sub-carriers to outer sub-carriers located at higher frequencies. Therefore, it was averaged over all OFDM symbols. We assumed a fixed frequency deviation between transmitter and receiver clock and make use of a simple adaptive scheme to estimate (or predict) an averaged phase slope $\Delta\bar{\theta}(n)$ from the instantaneous estimate $\Delta\theta(n)$:

$$\Delta\bar{\theta}(n) = [1 - \gamma] \cdot [\Delta\bar{\theta}(n-1) + \Delta\Delta\bar{\theta}(n-1)] + \gamma \cdot \Delta\theta(n), \tag{8.4}$$

$$\Delta\Delta\bar{\theta}(n) = [1 - \beta] \cdot \Delta\Delta\bar{\theta}(n-1) + \beta \cdot [\Delta\theta(n) - \Delta\theta(n-1)]. \tag{8.5}$$

After the phase slope is estimated, the receiver is able to correct the phase of pilot symbols. As described in Equation (8.2), these corrected pilot symbols are in the direction of CPE. The receiver performs a power-weighted sum of the pilots to

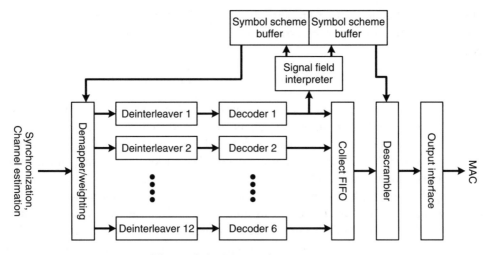

Figure 8.6   Data-path processor.

attain the CPE. After both phase parameters are estimated for OFDM symbols, data sub-carriers are processed. Before this, they are stored in FIFO buffer 3. The data equalizer, which is made up of a multiplier stage and a CORDIC stage, corrects the channel and estimated phase errors at the same time. The output of the receiver front-end becomes the phase-corrected and equalized constellation symbols on data sub-carriers. Depending on the data mode, they can be BPSK or any of the specified M-ary QAM constellations. Gray encoding is employed to minimize the bit error rate.

8.2.2   Receiver Back-End

Figure 8.6 is a block diagram of the receiver back-end which performs demapping/ weighting, deinterleaving, decoding and descrambling functions on the received constellation symbols from the receiver front-end. The initial soft-demapper is followed by a deinterleaver stage with 12 deinterleavers and a bank of 12 Viterbi decoders. After the signal field has been decoded, the contained data is analyzed in the signal field interpreter to calculate the current frame structure using our streaming algorithm. This algorithm is used to find the positions of the reference and termination symbols. The calculated data is then stored in symbol scheme buffers required for control of the receiver operation. All data transferred from decoders is stored in a data collector block. The purpose of this block is to unstream data and to perform descrambling. Finally, the processed data is delivered to the MAC through the output interface.

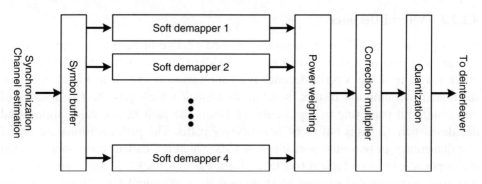

Figure 8.7   Demapper/weighting block.

8.2.2.1   Soft-Demapper

The implemented soft-demapper calculates bit metrics from received BPSK, QPSK, 16-QAM or 64-QAM symbols. These bit metrics are simplified log-likelihood ratios based on the derivations presented in [12], assuming Gaussian noise. A block diagram of this demapper is given in Figure 8.7. This processing block contains a symbol buffer to overcome possible processing delays caused by subsequent deinterleavers and Viterbi decoders. Four parallel demappers were needed to tackle four incoming parallel data streams coming from the channel estimator. The demapping process produces bit metrics, which are then weighted with the estimated power levels of the corresponding sub-carriers. This weighting is part of the bit metric calculation to account for different noise power contributions in different sub-carriers after equalization and therefore reflects the varying reliability of different sub-carriers received with higher and lower signal strength. After power weighting, it is necessary to scale the metrics with a constant factor for each modulation scheme. With this correction we are able to have quantization SNR degradation lower than 0.5 dB. Finally, data is quantized with 5-bit soft values.

8.2.2.2   Deinterleaver

An interleaver/deinterleaver works with address tables where the patterns for all transmission modes are stored. This approach facilitates rapid change of interleaving patterns corresponding data modes. Data shows up burstwise with soft bits from four sub-carriers in parallel. Since 64-QAM is the highest modulation scheme, up to 24 bits are stored per cycle in memory. The deinterleaver also performs depuncturing of the code stream by stuffing zeros. Therefore, a much more flexible puncturing scheme can be implemented, which is not limited to simple repetitive patterns.

8.2.2.3 Viterbi Decoder

In order to process 1 Gbps throughput, we used 12 Viterbi decoders, each of which can process data at a maximum rate of 100 Mbps or one output bit per cycle. A block diagram of the Viterbi decoder implementation is shown in Figure 8.8. The main area consumption results from the first block which gets the two metrics of one source bit per cycle as input data, performs the path metric calculations, and decides which 32 paths out of 64 are survivor paths. The path decisions are stored simultaneously in two dual port block RAMs which are called trellis memory. On the opposite side, two logical blocks perform the tracebacks. In the normal case, a traceback is accomplished over 96 steps to deliver 48 output bits. The design does not require any waiting cycles. Hence the throughput is half a source bit per cycle per traceback unit and the total throughput is one bit per cycle.

A significant reduction in complexity comes from the fact that no maximum path metric has to be calculated and tracebacks can always be performed starting from the zero state. This can be achieved when sufficiently long traceback length is used. In this case the path from any initial state will converge with very high probability

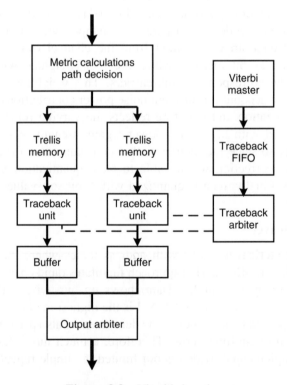

Figure 8.8 Viterbi decoder.

to the correct path. On the other hand, termination bits at the end of the stream ensure that the zero state is the final state, and with this *a priori* knowledge the last traceback can correctly start from this state.

The output bits during one traceback are assembled in reversed order. Therefore, they are stored in one shift register and copied to another shift register when all bits are read. Afterwards, they are read out by an output arbiter with 8 bits per cycle. The output arbiter reads the output of the buffers one after each other. It operates burstwise in that all bits belonging to a particular traceback have to be read out all together before moving on to the next unit.

In general, termination of the code stream for the purpose of decision feedback channel re-estimation can appear anywhere in the frame and the data stream may be continued afterwards. Therefore the traceback length will vary depending on the termination. The master block generates commands with complete information needed for a traceback. These commands are stored in a FIFO memory and read out by a traceback arbiter which serves the traceback units one after each other. The combined logic of the master and the traceback and output arbiters ensures that termination within the stream is properly handled. One traceback unit can never deliver output data before the other unit is finished with older data. The output bus must be at least two bits wide to ensure the correct operation.

8.3   OFDM Baseband Transmitter Architecture

The OFDM baseband transmitter is shown in Figure 8.9. Several blocks – the inverse FFT block, interleaver, and cyclic prefix insertion blocks – have almost identical architectures to those in the OFDM receiver, therefore detailed explanations of each block are not given here. The transmitter occupies 18 K flip-flops, 12 K slices, 41 multipliers and 111 block RAMs including demonstrator interface logic [13]. Data from a PC is acquired over the input control block. This block also calculates the cyclic redundancy check (CRC) for the signal field and initiates the calculation of the number of symbols. Input bit data is distributed to the 32-bit scrambler and then delivered to the symbol mapping block. This modulation

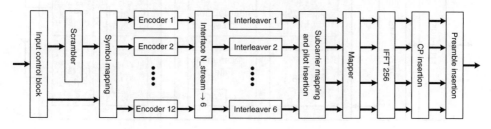

Figure 8.9   OFDM transmitter architecture.

mapping block distributes incoming data bytes with respect to modulation symbols. It also defines several related system parameters. If there is a termination, it will insert a dummy zero byte at the end of the symbol.

From this point, data flow is symbol oriented. Depending on the number of streams, data from this block is directed to between 1 and 12 convolutional encoders. Six interleavers running at 100 MHz was used to achieve the transfer data through-put of 1 Gsps. Therefore, we made an interface that redistributes data coming from 1–12 encoders into the fixed structure of six interleavers. Interleaved data is then mapped to the appropriate sub-carriers and pilot symbols are then inserted to pilot sub-carriers. To achieve 400 Msps effective throughput for OFDM modu-lation, we parallelized operation of the FFT with four individual 64-FFT blocks as described in Section 8.2. In the OFDM baseband hardware, each FFT also runs at the system frequency of 100 MHz. The tail part of data from FFT is duplicated and inserted as a cyclic prefix. Finally, the time-domain OFDM signals are multiplexed with stored preamble symbols and then transmitted to RF front-ends after digital-to-analog conversion.

8.4 60 GHz Link Demonstration

8.4.1 60 GHz OFDM Demonstrator Architecture

The architecture of an IHP 60 GHz OFDM demonstrator is illustrated in Figure 8.10. It consists of the developed OFDM baseband processor, the 60 GHz RF radios and MAC processor. These blocks were implemented separately on different hardware platforms.

The MAC block is made up of a processor for packet management and an FPGA hardware accelerator for time-critical calculation. Functionally, it first collects giga-bit Ethernet packets and then encapsulates them by adding a 8-byte header before the data and appending a 4-byte CRC. Afterward, the frames are delivered to the OFDM

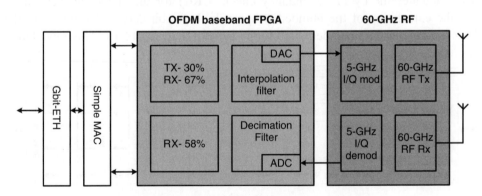

Figure 8.10   IHP 60 GHz OFDM demonstrator.

baseband part via a low-voltage differential signaling cable. The OFDM baseband parts were implemented with four Xilinx Virtex-II Pro FPGAs. Two FPGAs, which gives larger numbers of logics, were used for OFDM transmitter and receiver. The OFDM transmitter occupies 30% of one FPGA resource whereas the OFDM receiver uses 125%. The other smaller FPGA boards were utilized for baseband filtering, digital-to-analog and analog-to-digital of OFDM baseband signals. The outputs of two data-converters are connected to 60 GHz RF modules, specifically IF I/Q modulator and demodulator, with coaxial cables.

The 60 GHz RF front-end is based on a super-heterodyne transceiver architecture. An IF of 5 GHz is chosen because we used around 500 MHz bandwidth for one channel and this 5 GHz band is compatible with IEEE 802.11a wireless LAN standards. It is made up of four silicon chipsets which are a 60 GHz transmitter, a 60 GHz receiver, a 5 GHz I/Q modulator and a 5 GHz I/Q demodulator. They were fabricated in IHP in-house 0.25 μm SiGe BiCMOS technologies with an f_T of 200 GHz. To facilitate the field testing of fabricated chips, 60 GHz transmitter and receiver boards were designed with Rogers 3003 material 5 mm thick. A 60 GHz Vivaldi antenna was designed and integrated into these boards. It provides more than 8 dBi gain with the HPBW of 30° and 3 dB bandwidth larger than 7 GHz. All fabricated chips and boards are based on differential design. Figure 8.11 shows a 60 GHz transmitter board. The block diagrams for complete 60 GHz RF front-ends are shown in Figure 8.12. Detailed information on these chipsets and assembly boards can be found in [14].

8.4.2 Wireless Link Demonstration with 60 GHz Radio

The performance of the implemented OFDM baseband processors was first verified through a direct connection between baseband transmitter and receiver. Data

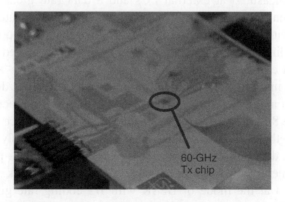

Figure 8.11 60 GHz transmitter board integrated with the transmitter chipset and Vivaldi antenna [4]. Reproduced by permission of © 2010 IEEE.

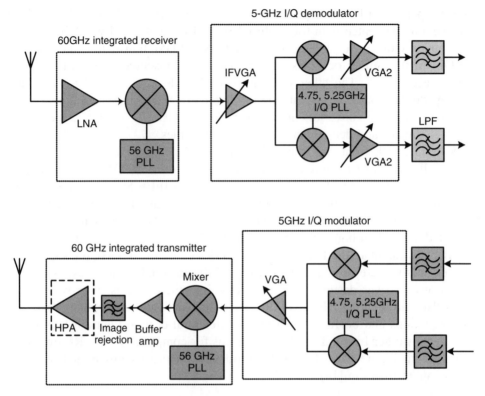

Figure 8.12 Block diagram of 60 GHz receiver and transmitter RF front-ends fabricated on IHP 0.25μm SiGe BiCMOS technology [4]. Reproduced by permission of © 2010 IEEE.

throughput was measured at external PCs and a data rate greater than 500 Mbps was obtained when only Ethernet links were applied. Bit error rate (BER) and FER were measured with a counter unit embedded on baseband components in order to characterize full-link performance. No frame error was observed when they were tested with direct connection links.

Afterward, the implemented basebands were connected to the developed 60 GHz analog front-ends and tested for 60 GHz link transmission in practical indoor environments. Figure 8.13(a) shows the test setup in an indoor office environment. As can be observed, the test environment did not exclude previously installed objects such as furniture, desks and cubicle walls, which introduce multi-path fading. Nevertheless, no frame error was observed over a 15 m 60 GHz link when the OFDM baseband operated at the mode of BPSK with 1/2 coding rate giving a 120 Mbps data rate [11]. The result confirmed that the developed OFDM baseband worked well and the included receiver functions are effective in the 60 GHz links influenced

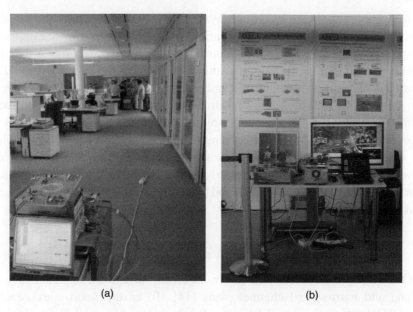

(a) (b)

Figure 8.13 (a) 60 GHz OFDM demonstrator setup in indoor office environments; (b) demonstrator setup for video transmission [4]. Reproduced by permission of © 2010 IEEE.

by practical multi-path fading and RF impairments. For transmission of the QPSK with 1/2 coding rate over 15 m distance, less than 10% FER was obtained, which is usually required in most of wireless LAN standards. Higher-order modulation such as 16-QAM was also possible while retaining less than 10% FER, but the maximum transmission distance was limited below 2 m. Previous simulations indicated that 16-QAM transmission with 1/2 coding would be possible over more than 3 m distance as long as an LOS link was maintained. The discrepancy is attributed to additional impairments arising from channels and RF non-idealities.

To demonstrate real-time video transmission, it is necessary to have bidirectional links for control signaling. Full duplex 60 GHz bidirectional links were realized by using two pairs of 60 GHz analog front-ends with frequency division. This was demonstrated at the International Conference on Communications 2009 (ICC 09) in Dresden, Germany.

8.5 Next-Generation OFDM Demonstrators for 60 GHz Wireless LAN Applications

The OFDM baseband processor developed within the WIGWAM project was designed to support a maximum of 1 Gbps with a channel bandwidth of 400 MHz for point-to-point applications. Nowadays, 60 GHz systems require far higher data throughput and wider coverage for various wireless applications. For example, uncompressed high-definition video transmission in 60 GHz requires a data rate of approximately 4 Gbps with 10 m transmission distance [15]. The IEEE 802.11ad task group is currently developing a standard for 60 GHz wireless LAN applications which specifies 1 Gbps throughput with extended coverage, supporting wireless LAN usage cases [16]. To meet such requirements imposed on 60 GHz systems, next-generation OFDM baseband is being developed with newly designed physical layers within the EASY-A project. The target application is multi-gigabit 60 GHz wireless LAN systems being addressed by the IEEE 802.11ad task group. This section briefly introduces the system and physical layer design of a next-generation OFDM demonstrator.

8.5.1 Channel Plan and RF Transceiver

New OFDM physical layers and RF transceivers have been designed to support wideband and narrowband channel plans [14]. To ensure good coexistence with legacy 60 GHz devices (e.g. IEEE 802.15.3c), the wideband channel plan is kept as defined in the 60 GHz standards. It is illustrated in Figure 8.14 as labeled W1 to W4. Between 57 and 66 GHz, four independent channels are defined, centred on 58.32, 60.48, 62.64 and 64.80 GHz (see Chapter 6). Each channel allows the maximum 2160 MHz channel bandwidth. To support the channel plan, we designed a PLL with a crystal oscillator of 19.2 MHz [17].

As illustrated in Figure 8.14, one wideband channel can be divided into three or four independent narrowband channels as long as centre frequencies can be generated from the same PLL. This narrowband channel plan is particularly suited for wireless LAN systems where one access point needs to accommodate a larger number of simultaneous user devices. The division into four narrowband channels makes it possible to use 12 channels over a 7 GHz bandwidth, which is almost comparable to IEEE 802.11a standards supporting 12 channels in the 5 GHz band.

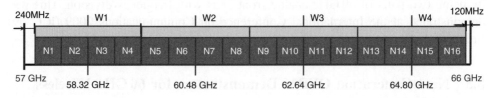

Figure 8.14 Wideband and narrowband channel plans for next-generation OFDM baseband.

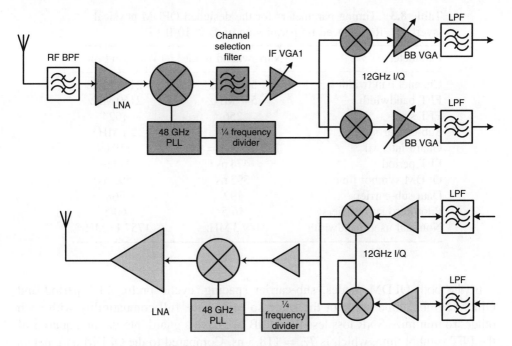

Figure 8.15   60 GHz sliding IF architecture.

Figure 8.15 shows a block diagram for the next-generation 60 GHz transceiver architecture. It is based on a sliding IF architecture where a 12 GHz IF band is generated from 48 GHz LO by frequency division. Because it eliminates the use of an additional crystal oscillator and PLL for IF conversion, lower power consumption and easier integration can be expected while maintaining the merits of a super-heterodyne scheme. We also leave out an image rejection filter by incorporating higher IF (12 GHz) with the frequency-selective response of 60 GHz components.

8.5.2   Next-Generation Multi-Gbps OFDM Physical Layers

Two OFDM physical layers are designed for wideband and narrowband channel plans and will be implemented on FPGA platform [18]. The sampling frequencies for wideband and narrowband OFDM modes are 2160 and 540 MHz, respectively. Both frequencies can easily be generated from the PLL architecture used for the generation of RF center frequencies. For quick hardware implementation, we decided the sampling frequency for the wideband scheme considering commercially available ADCs supporting higher than 6-bit resolution. Table 8.3 summarizes timing parameters for both wideband and narrowband OFDM modes.

Table 8.3 Timing parameters for the designed OFDM physical layers [4]. Reproduced by permission of © 2010 IEEE

Parameter	Narrowband mode	Wideband mode
Channel bandwidth	540 MHz	2160 MHz
FFT bandwidth	540 MHz	2160 MHz
FFT size	256	1024
Sub-carrier spacing	2.1 MHz	2.1 MHz
Guard interval	119 ns ($N_G = 64$)	119 ns
FFT period	474 ns	474 ns
OFDM symbol time	593 ns	593 ns
Data sub-carriers	192	768
Pilot/Zero sub-carriers	16/5	60/5
Nominal used bandwidth	449.3 MHz	1757.11 MHz

In the both OFDM modes, sub-carrier spacing, cyclic prefix, FFT period and OFDM symbol time are kept the same so as to have full compatibility with each other. To minimize SNR loss less than 1 dB, we have a guard interval of a quarter of the FFT symbol time, which is $T_G = 118.5$ ns. Compared to the OFDM parameters previously developed, the new OFDM physical layer is less susceptible to phase noise but a little more sensitive to channel delay spread. Each mode supports different transmission rates with the modulations from BPSK to 64-QAM with various coding rates.

To obtain higher coding gain than the previous one utilizing convolutional code, two different channel coding approaches are considered. One is to apply Reed–Solomon channel coding as the outer code since it improves the coding gain of the convolutional inner code without sacrificing hardware complexity and code rate significantly. The Reed–Solomon (255, 253) code specified in the IEEE 802.15.3c standard is used for this purpose [1]. The other approach is to employ a low-density parity check (LDPC) code due to its its higher coding gain. We use the (768, 384) LDPC code specified in IEEE 802.16e standard because of its block length fitted to the number of OFDM data sub-carriers and its feasibility to implement on small chip size [19].

8.5.3 Performance Evaluation with 60 GHz NLOS Channel and 60 GHz Phase Noise Models

The main object of performance evaluation is to optimize a designed physical layer block by block before starting FPGA implementation. To have more accurate estimation close to practical results, full link simulation included 60 GHz NLOS channel and 60 GHz phase noise models which were believed to be the most dominant sources of impairments. There are several 60 GHz channel models available in the IEEE 802.15.3c task group. Among them, we used the NLOS residential channel model (corresponding to CM2.3) because it was the mandatory channel model used in the proposal evaluation [20]. The 30° HPBW antenna was used for spatial filtering in both transmitter and receiver.

For simulation, hundreds of 2048-byte frames were transmitted and the mean 90% BER link success probability was computed as described [20]. The energy in guard intervals, pilot carriers and preambles was taken into account for practical FER simulation. We used all receiver functions, including frame synchronization, channel estimation/equalization and common phase-error corrections using defined preambles and pilot carriers [A]. The performance of a wideband OFDM physical layer for 60 GHz residential NLOS model is presented in Figure 8.16. No error floor is observed in the required 10% FER. We also compared it with practical case where phase noise is applied. The single sideband phase noise value of $L_{SSB} = -93$ dBc/Hz at 1 MHz offset was applied for simulation. We found that a frame error rate of 10% was achieved at an E_b/N_0 of 13.2 dB. Less than 0.5 dB degradation was observed for QPSK and 16-QAM modulations.

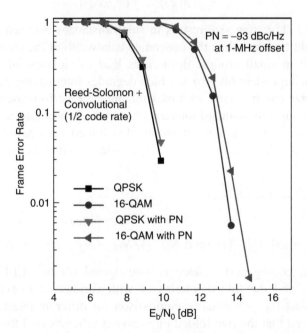

Figure 8.16 Performance evaluation of the designed wideband OFDM physical layer under 60 GHz NLOS residential channel and 60 GHz phase noise models [4]. Reproduced by permission of © 2010 IEEE.

From the previous demonstrator, it was also noticed that interleaving schemes improve performance under NLOS channel environments. For convolutional codes, tone interleaving is required to spread adjacent coded bits over the whole spectrum. In addition, adjacent coded bits should have alternating bit reliability when they are mapped on M-ary QAM symbols. The standard permutation defined in the IEEE 802.11a standard meets both requirements. The permutation rule is given below in slightly different notation:

$$N_{block} = N_D \cdot m, \quad N_{round} = N_D/\Delta_{SC},$$

$$\Delta_{bit} = m \cdot \Delta_{SC}, \quad s = \max(m/2, 1),$$

$$p(k) = \Delta_{bit} \cdot (k \bmod N_{round}) + \lfloor k/N_{round} \rfloor,$$

$$q(p) = s \cdot \lfloor p/s \rfloor + ((p + N_{block} - \lfloor N_{round} \cdot p/N_{block} \rfloor) \bmod s).$$

The first permutation rule completely fills N_{round} sub-carriers which lie on a grid with the step size of Δ_{SC} before the next of such a cycle group is processed in the same way. The only free parameter of this interleaver is the step size of Δ_{SC}. We therefore write the permutation as a function of the input index and step size of Δ_{SC}:

$$y = q(p(x)) \equiv f(x, \Delta_{SC}),$$

Choosing a very short step size results in small distances between adjacent coded bits, which might be lower than the coherence bandwidth of the channel. Very large step sizes result in small groups. Both cases lead to violation of the assumption of statistically independent bit errors, which degrades interleaving performance. An optimum step size can be found with maximizing code performance. We added the third permutation to this standard interleaver, which enables longer groups without the penalty of a small step size. It is called a folded interleaver. An additional free parameter, the folding factor N_{fold}, provides a wider range of permutation schemes [21]:

$$y = f(x, \Delta_{SC}/N_{fold}),$$

$$N_{block} = N_D \cdot m/N_{fold}, \quad N_1 = m \cdot \Delta_{SC}/N_{fold}, \quad N_2 = m \cdot \Delta_{SC},$$

$$z = y \bmod N_1 + \lfloor (y \bmod N_{block})/N_1 \rfloor \cdot N_2 + N_1 \cdot \lfloor y/N_{block} \rfloor.$$

Convolutional coding performance is investigated for the IEEE 802.11a standard interleaver and the folded interleaver with various interleaving parameters. Figure 8.17 shows the performance comparison for different interleaving parameters. It is observed that the two folded interleaving schemes slightly outperform the standard interleaver with the best parameters. The performance gain is approximately 0.5 dB at a BER of 10^{-5} and it should be noted that it comes at no additional expense.

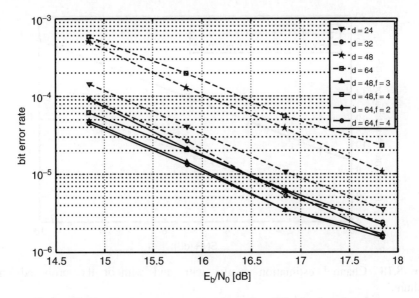

Figure 8.17 Performance comparison for different interleavers.

To achieve higher throughput, the preamble length is shortened from 11 OFDM symbols in the previous case to 6.6 symbols. This was possible with the help of the improved channel estimation scheme which makes it possible to reduce the length of preamble from 4 symbols to 2. In 60 GHz channels, the channel impulse response length is usually much shorter than the FFT length. The proposed scheme is to filter out the estimated channel impulse responses far away from actual channel impulse responses, which are likely to be just noise in channel estimation. This potentially gives channel smoothing functions in the frequency domain. The initial estimates for the channel coefficients H_n are obtained by averaging two long preamble symbols. From these estimates, the mean phase $\Delta\phi$ increase from one sub-carrier to next is estimated with a phase unwrapping scheme [21]. Then channel coefficients are rotated to attain zero phase progression. In this way, the corresponding impulse response is shifted in the time domain, so that the main energy impulse response can be located at zero time:

$$G_n = H_n \cdot \exp(-j\,\Delta\phi \cdot n).$$

After this rotation, a filter convolution in the frequency domain is applied, which provides low-pass filter function in the time domain. The effect of this filter B_n is to attenuate the aforementioned noise components far away from the channel impulse response:

$$K_n = G_n \otimes B_n.$$

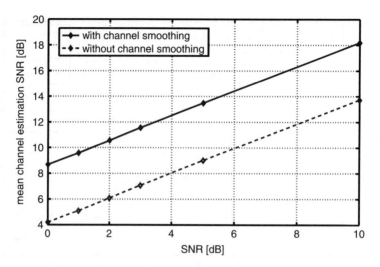

Figure 8.18 Channel estimation results with and without the proposed channel smoothing.

The improvement in channel estimation performance for the 60 GHz NLOS residential model is shown in Figure 8.18. More than 4 dB improvement in the required SNR is obtained.

References

[1] IEEE 802.15.3c (2009) *Part 15.3: Wireless medium access control (MAC) and physical layer (PHY) specification for high rate wireless personal area networks (WPANs): Amendment 2: Millimeter-wave based alternative physical layer extension*.

[2] http://www.wigwam-project.de

[3] http://www.easy-a.de/

[4] Choi, C.-S., Grass, E., Piz, M., Ehrig, M., Marinkovic, M., Kraemer, R. and Scheytt, C. (2010) 60-GHz OFDM systems for multi-gigabit wireless LAN applications. *IEEE Consumer Communications and Networking Conference*, Las Vegas.

[5] Heinemann, B., Barth, R., Knoll, D., Ruecker, H., Tillack, B. and Winkler, W. (2007) High performance BiCMOS technologies without epitaxially-buried subcollectors and deep trenches. *Semiconductor Science and Technology*, **22**, 153–157.

[6] Peter, M., Keusgen, W., Kortke, A. and Schirrmacher, M. (2007) Measurement and analysis of the 60-GHz in-vehicular broadband radio channel. *66th IEEE Vehicular Technology Conference*, Fall.

[7] Sawada, H., Shoji, Y., Choi, C.-S., Sato, K., Funada, R., Harada, H., Kato, S., Umehira, M. and Ogawa, H. (2006) LOS office channel model based on TSV model. IEEE 802.15-06-0377-01-003c, IEEE 802.15.3c, 2006.

[8] Xu, H., Kukshya V. and Rappaport, T. S. (2002) Spatial and temporal characteristics of 60-GHz indoor channels. *IEEE Journal of Selected Areas in Communications*, **20**(3), 620–630.

[9] Manabe, T., Miura, Y. and Ihara, T. (1996) Effects of antenna directivity and polarization on indoor multipath propagation characteristics at 60-GHz. *IEEE Journal of Selected Areas in Communications*, 14(3), 441–448.

[10] Piz, M. and Grass, E. (2007) A synchronization scheme for OFDM-based 60 GHz WPANs. *IEEE International Symposium on Personal, Indoor and Mobile Radio Communications*, Athens, September.

[11] Piz, M., Krstic, M., Ehrig, M. and Grass, E. (2009) An OFDM baseband receiver for short-range communication at 60 GHz. *IEEE International Symposium on Circuits and Systems*, Taipei, May.

[12] Tosato, F. and Bisaglia, P. (2002) Simplified soft-output demapper for binary interleaved COFDM with application to HIPERLAN/2. *IEEE International Conference on Communications*, New York, April/May.

[13] Krstic, M., Piz, M. and Grass, E. (2008) 60 GHz datapath processor for 1 Gbit/s. *VLSI-SOC*, October.

[14] Choi, C.-S., Grass, E., Herzel, F., Piz, M., Schmalz, K., Sun, Y., et al. (2008) 60 GHz OFDM hardware demonstrators in SiGe BiCMOS: state-of-the-art and future development. *IEEE International Symposium on Personal, Indoor and Mobile Radio Communications*, Cannes, September.

[15] Wireless HD Consortium. http://www.wirelesshd.org

[16] IEEE 802.11ad task group. http://www.ieee802.org/11/Reports/tgad_update.htm

[17] Herzel, F., Choi, C.-S. and Grass, E. (2009) Frequency synthesis for 60-GHz OFDM transceiver. *EuWiT 2008*, Amsterdam, October.

[18] Choi, C.-S., Piz, M. and Grass, E. (2009) Performance evaluation of Gbps OFDM PHY layers for 60-GHz wireless LAN applications. *IEEE International Symposium on Personal, Indoor and Mobile Radio Communications*, Tokyo, September.

[19] IEEE 802.16e (2005) *Part 16: Air interface for fixed and mobile broadband wireless access systems: Amendment 2: Physical and medium access control layers for combined fixed and mobile operation in licensed bands*.

[20] Seyedi, A. et al. (2008) TG3c system requirement. IEEE 802.15-07-0689-00-003c, IEEE 802.15.3c, 2008.

[21] Piz, M., Grass, E., Marinkovic, M. and Kraemer, R. (2009) Next-generation wireless OFDM systems for 60-GHz short-range communication at a data rate of 2.6 Gbps. *OFDM Workshop*, Hamburg, September.

9

Medium Access Control Design

Harkirat Singh

The medium access control (MAC) protocols control the access rights of the shared medium. In a half-duplex wireless channel, a node cannot sense the channel while transmitting a frame. Therefore, an efficient MAC protocol becomes necessary. Since collisions cannot be detected by the transmitter, a typical MAC scheme attempts to decrease the probability of a collision using collision avoidance principles. For instance, in IEEE 802.11, a device wishing to access the channel first senses the medium to determine if it is idle. If so, the device waits an additional, randomly selected, period of time and then transmits if the medium is still free. However, if it is determined that the channel is busy, the device may enter into a random backoff period and wait until transmission stops. This prevents multiple devices seizing the medium immediately after completion of preceding transmission. Such carrier sensing based MAC schemes rely on an omnidirectional antenna that receives and transmits radio frequency (RF) energy equally in all directions. While omnidirectional antenna patterns assist an efficient medium access, they create undesired interference at some of the other devices.

The 60 GHz band provides an extremely short wavelength of 5 millimeters that reduces the antenna size significantly. Now it becomes feasible to mount a large array of antenna elements (or directional antennas) on a node. Directional antennas provide some advantages over omnidirectional antennas. For example, directional antennas radiate more energy in smaller zones, giving some benefits in terms of reducing unwanted interference provided that the intended power is directed towards the desired users. This also helps in improving the signal-to-noise ratio (SNR) at the intended receiver. In addition, directional antennas increase network throughput via spatial reuse, thus allowing more simultaneous transmissions when compared with

60 GHz Technology for Gbps WLAN and WPAN: From Theory to Practice
Edited by Su-Khiong (SK) Yong, Pengfei Xia and Alberto Valdes Garcia
© 2011 John Wiley & Sons, Ltd

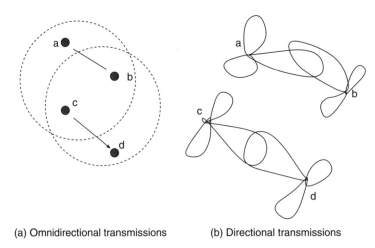

(a) Omnidirectional transmissions   (b) Directional transmissions

Figure 9.1 Omnidirectional and directional antennas [1]. Reproduced by permission © 2008 IEEE.

omnidirectional antennas. Figure 9.1 illustrates a scenario of the increased spatial reuse capability provided by directional antennas. Nodes a and c have packets for nodes b and d, respectively. If a and c transmit simultaneously using omnidirectional antennas then the frames will collide, and both the nodes will perform a random backoff and try to retransmit independently in the future. However, using directional antennas, a and c can successfully transmit data packets simultaneously. Nevertheless, directional antennas brings new challenges for the MAC design. For example, nodes with directional antennas may not be able to receive in all directions, and therefore, may miss control and/or data frames from some other directions.

In this chapter, the challenges of MAC layer design in the presence of directional antennas are discussed. In order to support multi-gigabit data rate at the MAC service access point, MAC efficiency should be targeted at 80% and beyond. A number of techniques to improve the MAC service access point efficiency such as a large packet size (of the order of hundreds of kilobytes), data aggregation, block-ACK and automatic repeat request (ARQ), are addressed. Finally, MAC design considerations for supporting uncompressed video are detailed.

9.1 Design Issues in the Use of Directional Antennas

In this section we enumerate issues that arise at the MAC layer due to the use of directional antennas. One of the prominent artifacts that directional antennas creates is *deafness*, whereby neighboring nodes are hidden from each other not only due to distance separation, but also because of differences in the orientation of directional antenna beams. Therefore, device discovery becomes challenging in the presence of

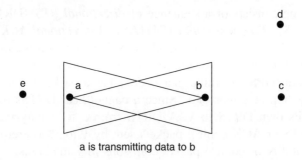

Figure 9.2   Deafness and hidden terminal problems.

directional (or beamformed) transmissions. Identifying the neighbors is necessary before two devices can communicate with each other.

Figure 9.2 illustrates some of the hidden terminal and deafness problems that may occur in the presence of directional antennas. Here, node e is a one-hop neighbor of node a, and nodes c and d are one-hop neighbors of node b. Assume that node a is sending data to node b. Consider the following cases:

- c wants to transmit to b. In this case, c will send a request to send (RTS) frame (as in IEEE 802.11) to b since c senses no activity on the channel. Given that b has a linear antenna array, the beam pattern at b is oriented towards a and prohibits b receiving from c. Assuming that a is sending a large amount of data to b, c may make multiple retransmission attempts that cause a large backoff window at c. Suppose that the transmission from a to b finishes at a certain point. Node c would have started transmitting to b immediately, had it known that b would be free to receive new packets. However, because c does not know this by the current network protocol, c has to wait and decrement the backoff unnecessarily. Thus, directional transmissions not only create unwanted interference at b from c, but also waste channel resources in some cases. These problems do not arise when the nodes use omnidirectional antennas.
- d wants to transmit to b. This case is similar to when c wants to transmit to b.
- Nodes c and d are hidden from a due to directional transmissions and e is hidden from b due to distance separation.

Ko et al. [2] solve the deafness problem by using a combination of *omnidirectional and directional* RTS and clear to send (CTS). They assume that the gain of the directional antenna equals that of the omnidirectional antenna. The antenna model is assumed to be a sectored antenna which can transmit in $90°$ sectors only. Hence, such an antenna system requires four antenna elements to cover all directions. Two different schemes are evaluated.

The first scheme consists of a sequence of *directional RTS* (DRTS), *omnidirectional CTS* (OCTS), *directional data* (DDTA) and *directional ACK* (DACK). The control packets (RTS/CTS) contain the physical location information of the sender and the receiver. The physical location information may be obtained by the global positioning system (GPS). Assume that node x has a packet for node y, and node a is the previous-hop neighbor of x. Since a cannot hear DRTSs from x, hence, it is free to send its own DRTS to x. A DRTS from a to x may interfere with the reception of OCTS or ACK control packets sent by y to x. Therefore, this scheme has a high probability of control packet collisions in some cases.

The second scheme alleviates the problem of a higher number of collisions by transmitting a combination of DRTS and omnidirectional RTS (ORTS). A node keeps track of its blocked directional antennas, which remain *blocked* for the duration of the packet received on that particular antenna sector – the duration of transfer is included in each control packet such as RTS and CTS (as in IEEE 802.11). Hence, x will send an ORTS if none of its directional antennas at x are blocked; otherwise, x will send a DRTS provided that the desired directional antenna is not blocked. The first scheme allows multiple simultaneous transmissions at the cost of higher probability of collisions of control packets; the second scheme reduces the possibility of such collisions at the cost of reduced spatial reuse of the channel.

Nasipuri et al. [3] propose an MAC protocol that uses directional antennas where mobile nodes do not have any location information. Each node is equipped with M directional antenna elements. Each of the antenna elements has a conical pattern, spanning an angle of $2\pi/M$ radians. The M antennas at each node are fixed with non-overlapping beam directions so as to collectively span the entire plane. The MAC protocol is assumed to be capable of switching any one or all the antennas to active or passive modes. In this work the authors assume that all the antennas have the same gain. The protocol uses omnidirectional transmission of the RTS/CTS control packets. The receiver uses *selection diversity* – that is, the receiver uses the signal from the antenna that is receiving maximum signal strength. The receiver records the antenna that received the *maximum signal power*, and thus the receiver also identifies the direction of the maximum signal power. The sender uses this information to directionally transmit the data packet followed by directional exchange of the acknowledgement (ACK). The proposed solution alleviates the deafness problem at the cost of poor spatial reuse.

In order to efficiently perform collision avoidance, the IEEE 802.11 (2007) protocol [4] does both virtual and physical carrier sensing before transmitting a packet. Each transmitted packet has a duration field that indicates how long the remaining transmission will be. So, the overhearing nodes can remain silent for this duration. The overhearing node records this value in a variable called the network allocation vector (NAV). When a node has data to send, it first looks at the NAV. If its value is not zero, the node determines that the medium is busy. This is called virtual carrier sensing. Takai et al. [5] extend the NAV usage to the environment where nodes are equipped with directional antennas. The new protocol is called directional virtual carrier sensing (DVCS). A node caches direction of arrival (DOA) information based on signals received, and nodes remain in promiscuous mode to cache signals. In DVCS a node transmits four directional RTSs and the remaining three as omnidirectional RTSs if there is no response to the directional RTSs. The proposed scheme achieves good spatial reuse. Choudhury et al. [6] propose directional MAC (DMAC), which also aims to maximize the spatial reuse, and suggest that all MAC layer operations should be performed directionally. However, these protocols do not solve the deafness problem.

ElBatt et al. [7] introduce the interesting idea of blocked and unblocked beams. They propose to modify RTS/CTS packets by including the beam index on which the ensuing data packet will be sent. The RTS/CTS packets are transmitted on all unblocked beams. A node lying in the coverage of a blocked beam would be left unaware of the attempted reservation. This leads to a deafness problem. To solve this problem they send another set of RTS/CTS packets on blocked beams on a different frequency than the one used for sensing reservation frames on unblocked channels. Thus each node needs to be equipped with two radios to overcome with deafness problem. Directional hidden terminals still remain a problem in [7].

Korakis et al. [8] solve the deafness problem by emulating an omnidirectional RTS by transmitting it sequentially multiple times using directional beams. If a node requires M beams to cover $360°$ then the node will transmit the RTS control frame M times. The destination node replies with a single directional CTS frame. Note that in a scenario where the chances of spatial reuse are limited (e.g. linear network), such a scheme will not be of much help. Moreover, in such scenarios increasing the number of antenna elements will reduce the beamwidth and will negatively affect the throughput performance.

Choudhury and Vaidya [9] address the deafness problem by transmitting an out-of-band tone omnidirectionally following the completion of data transmission between two nodes. Out-of-band tones act as a corrective method for nodes that suffer from deafness by terminating the backoff if the address of the tone originator matches with the intended receiver that caused the node to invoke repeated backoffs. The proposed scheme does not solve the directional hidden terminal problem in the presence of missed RTS/CTS.

In summary, directional transmissions help in boosting the overall system throughput, but result in deafness and directional hidden terminal problems that, if not corrected, can easily offset the benefits of directional transmissions. In the next section, we present changes to the IEEE 802.15.3 MAC to deal with directional 60 GHz transmissions.

9.2   IEEE 802.15.3c MAC for 60 GHz

IEEE 802.15.3 (2005) [10] provides a time division multiple access (TDMA) based MAC protocol for a wireless personal area network (WPAN) that operates in relatively short distances, about 10 meters. In IEEE 802.15.3, a piconet coordinator (PNC) periodically sends beacons that disseminate various details about the superframe such as beacon interval, contention access period (CAP), and contention free period, i.e. channel time allocation (CTA). The beacon also helps in maintaining time synchronization in the WPAN. In contrast to contention based MAC protocols, TDMA based protocols are well suited for 60 GHz directional transmission to overcome the deafness and neighbor discover issues presented in Section 9.1. In this section we investigate the changes made by IEEE 802.15.3c (2009) [11] to the IEEE 802.15.3 MAC in order to deal with 60 GHz directional transmission.

9.2.1   Neighbor Discovery

IEEE 802.15.3c (2009) [11] provides a simple neighbor discovery for PNC based WPANs. Suppose that a PNC has P_{TX} transmit antennas and P_{RX} receive antennas such that it can transmit and receive in P_{TX} and P_{RX} directions, respectively. Similarly, a device[1] has D_{TX} transmit antennas and D_{RX} receive antennas. The PNC transmits quasi-omni (Q-omni) beacons P_{TX} times, sending each repetition with one of the P_{TX} antennas to enable devices located in different angular directions to join the same piconet, as shown in Figure 9.3. The device listens to the Q-omni beacons in all D_{RX} directions to find the best and the second best direction pair based on the link quality indicator, for example SNR. The device compares at least $P_{TX} \times D_{RX}$ pairs before deciding the best and second best pair. Afterwards, the device informs the PNC about its best transmit direction in the association command transmitted during the association sub-CAP (S-CAP). Notice that in IEEE 802.15.3 a single CAP is used for omnidirectional transmissions and receptions, but IEEE 802.15.3c divides the single CAP into multiple S-CAPs, each aligned with P_{RX} receive directions at the PNC. Therefore, at least P_{RX} S-CAPs are required for the PNC to discover all possible neighbors. For the symmetric antenna system (SAS) case the transmit direction at the PNC (or at a device) is used for signal reception as well. In such scenarios $P_{TX} = P_{RX}$ and $D_{TX} = D_{RX}$. A device, upon receiving the best Q-omni beacon from the PNC in the $\mathcal{D}_{RX}$ direction, replies with a response during an S-CAP with the same index as the PNC's optimal transmit beam $\mathcal{P}_{TX}$ for the Q-omni beacon. However, in the asymmetric antenna system (AAS) case, where transmit directions are different from receive directions ($P_{TX} \neq P_{RX}$), a device does not know the optimum receive direction at the PNC even after finding the optimal

---

[1]The term *device* is used in IEEE 802.15 and *station* is used in IEEE 802.11; both are used interchangeably in this chapter.

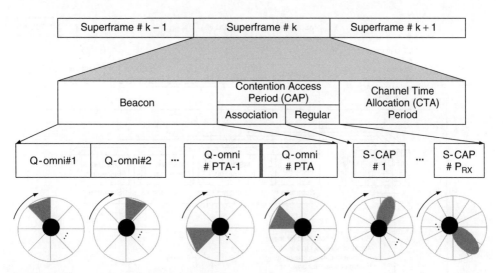

Figure 9.3 The Q-omni beacon and contention access period in 802.15.3c for PNC device discovery [11]. Reproduced by permission © IEEE.

Q-omni beacon (i.e. $\mathcal{P}_{TX}$). Therefore, a device needs to transmit the association request command at different D_{TX} antennas during the association S-CAP until the PNC replies with the association response command.

The amount of time required to complete one round of neighbor discovery scales with the number of antennas at the PNC as well as the maximum number of antennas for the devices. A two-stage neighbor discovery could be used to optimize the neighbor discovery period [11] that is discussed in Section 4.2. The first stage includes coarse-level neighbor discovery and the second stage performs a fine-level discovery within the discovered coarse-level direction. Since the coarse-level discovery process uses a wider beamwidth antenna pattern (or a wider directional sector), the amount of time required to cover $360°$ space is far less than when a narrower beamwidth is used.

9.2.2 Aggregation and Block-ACK

IEEE 802.11n provides two kinds of aggregation schemes. In the aggregate MAC service data unit (A-MSDU) scheme, multiple MSDUs are aggregated in one MAC frame, typically at the top of the MAC layer. In the aggregate MAC protocol data unit (A-MPDU) scheme, multiple MPDUs are aggregated in one MAC frame at the bottom of the MAC layer. The aggregated frame is forwarded to the physical layer (PHY) for transmission. One potential drawback of these aggregation schemes is that individual MSDUs cannot be retransmitted since the minimum retransmission

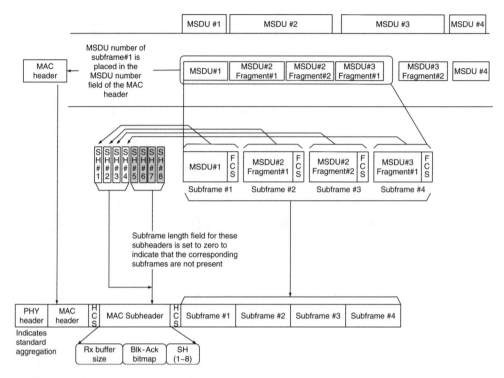

Figure 9.4   Standard aggregation [11]. Reproduced by permission © 2009 IEEE.

unit is a single MPDU. Another significant limitation is that MSDUs in the aggregated frame cannot be coded differently using a different modulation and coding sequence (MCS).

The aggregation scheme in IEEE 802.15.3c overcomes the aforementioned drawbacks by allowing retransmissions of individual MSDUs and allowing different coding rates for aggregated MSDUs. IEEE 802.15.3c provides three aggregation modes, namely, standard aggregation, low-latency aggregation, and audio video (AV) aggregation presented next.

9.2.2.1   Standard Aggregation

This may be used for aggregating high speed data and AV. Figure 9.4 illustrates the standard aggregation at the source. An MSDU may be mapped to one or more subframe payloads. A subframe header (SH) is created for each valid subframe payload. The SH contains the necessary information for the destination device to accurately retrieve the original data by deaggregating the received MPDU. The MSDU number of the first subframe is placed in the MSDU number field of the MAC header. The

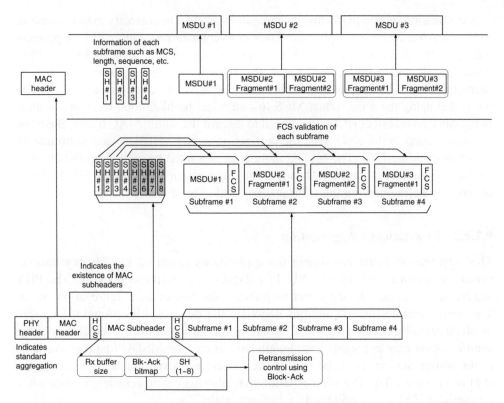

Figure 9.5 Standard deaggregation [11]. Reproduced by permission © 2009 IEEE.

MSDU number for a subsequent subframe is represented by its offset relative to the first subframe. Besides MCS, the SH field also includes the subframe length, MSDU fragmentation indication, retransmission indication, and frame checksum (FCS) flag. Since at most eight subframes can be aggregated in one MPDU, the eight SHs are combined together and placed in the MAC subheader as shown in Figure 9.4. For an MPDU containing less than eight subframes, the subframe length in the corresponding SH is set to zero to indicate that the subframe payload is not present. Standard aggregation supports both unidirectional and bidirectional data transmission by aggregating ACK information with data. For unidirectional transmission, the destination device upon receiving an aggregated MPDU replies with the block-ACK appropriately set in the MAC subheader (Figure 9.5) to indicate the status of each subframe. Since no data is included in the aggregated frame for unidirectional transmission, the subframe length field in all eight SHs is set to zero to indicate that no data is aggregated. For bidirectional data transmission, the destination device aggregates the block-ACK with data in subframe payloads, if any.

A destination device performs deaggregation process to correctly extract received MSDUs. After receiving a physical layer convergence procedure (PLCP) protocol data unit, the destination device checks the aggregation type indicated in the PHY header as shown in Figure 9.5. Later, the HCS fields are validated to verify the correctness of MAC header and subheader. The MAC header and subheader fields are coded using the most robust MCS to minimize the likelihood of errors in these fields since an incorrect HCS would need to discard the whole MAC frame. Based on the information in the eight SHs, the destination device validates each subframe by the corresponding FCS. The MSDU number is generated by adding up the offset to the MSDU number in the MAC header. The fragmentation field in the SH facilitates in properly defragmenting the subframe payload(s) to re-create the original MSDU.

9.2.2.2 Low-Latency Aggregation

This aggregation mode is suitable for applications requiring low-latency bidirectional communication such as USB, PCI Express, and wireless docking. The PHY header at the source device is set to indicate the low-latency aggregation mode. The source and destination stations negotiate the desired bidirectional CTA relinquish duration by exchanging the CTA relinquish duration information element. The source station may aggregate zero-length data in case no MSDU (data) is available at the source station from the end of the current CTA relinquished period to the start of the next CTA. The source station may also aggregate zero-length data when no periodic MSDU is available in a bidirectional CTA.

As shown in Figure 9.6, the source MAC layer attaches an MSDU SH to each MSDU received from the higher layer. The MSDU SH field contains a MSDU number that uniquely identifies a MSDU in the series of MSDUs delivered to the destination device. The subframe length field contains the length of a subframe. If periodic data is not present then a zero subframe length idle data is aggregated. The zero-length MSDU sequence number is assigned to the most recent acknowledged sequence number at the source. The MSDU response number field in the MAC SH indicates the first MSDU sequence number of the transmitted block-ACK bitmap field. The MSDU response number field is used to generate the offset for the block-ACK bitmap field. The MSDU request number field indicates the most recent MSDU sequence number acknowledged at the transmitter.

A destination device validates the MSDU SH HCS field for each MSDU SH. If the HCS is not valid, the destination device then searches for a valid MSDU SH HCS until one is found by matching a valid MSDU HCS. Once a match is found, the destination device then checks the validity of the FCS field computed over the expected subframe length payload. If the MSDU is incorrectly received then the corresponding bit is set to zero in the next transmitted block-ACK bitmap field according to its order offset from the MSDU response number field as shown in

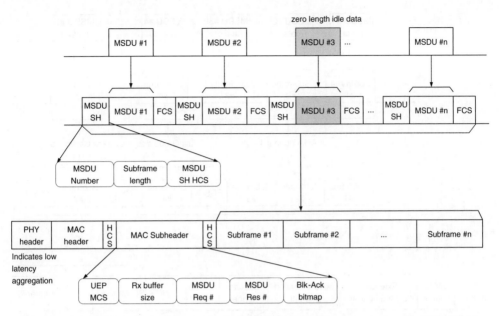

Figure 9.6 Low-latency aggregation [11]. Reproduced by permission © 2009 IEEE.

Figure 9.6. The source and destination devices interchange roles of data transmission and reception during the CTA every bidirectional CTA relinquish duration or less by using the IEEE 802.15.3 CTA relinquish method [10] that allows another device to transmit data in a given CTA. In the low-latency mode at most 256 subframes may be aggregated in one aggregated frame.

9.2.2.3   AV Aggregation

The AV aggregation for the high-rate PHY (HRP) in the AV OFDM is optimized to carry uncompressed audio and video in an efficient manner. At most seven subframes may be aggregated in one aggregated frame. As shown in Figure 9.7, the extended MAC header includes the MAC extension header, security header, and video header. The MAC extension header field includes the type and ACK group fields. The type field indicates the type of data in the subframe. The valid values for the type field are: MAC command, Data, Audio, and Video. The ACK group is used to map seven subframes to five ACK bits. Therefore, at most five FCS fields are added to seven subframes. A new ACK group is started when the bit in the ACK group corresponding to the subframe is set to zero. However, the bit for the subframe is set to one if the subframe is part of the previous ACK group. Thus, the first bit in the ACK group field, corresponding to the first subframe, is always set to zero. Also, the total number of bits set to zero can be at most five.

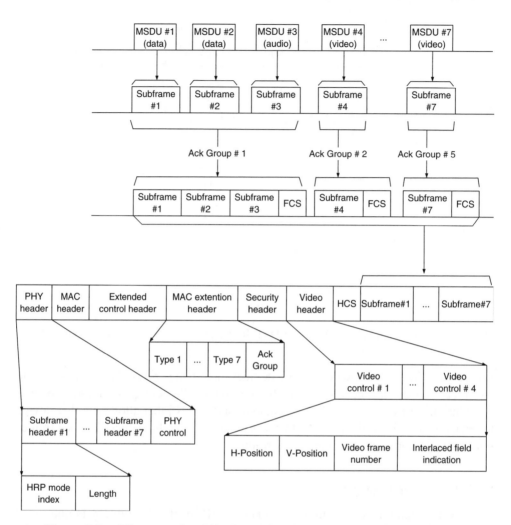

Figure 9.7   AV aggregation [11]. Reproduced by permission © 2009 IEEE.

Figure 9.7 presents a case where subframes 1–3 are mapped to the first ACK group. The remaining four subframes (4–7) are each uniquely mapped to each of the four ACK groups, respectively. In this example, the ACK group field is set to "0110000". A checksum is computed for each ACK group and stored in the FCS field.

In AV aggregation, video subframes are neither aggregated nor fragmented; it is therefore possible to have one-to-one mapping between video MSDUs and the corresponding subframes. For each video MSDU, the video header contains video related information. At most four video subframes may be aggregated in one aggregated MAC frame since the total number of video control fields in the video header is fixed and set to four as shown in Figure 9.7. The H-position and V-position fields contain the horizontal and vertical positions (in the video frame) of the first pixel in the subframe, respectively. The video frame number field contains a counter that keeps track of the video frame to which the pixels in the subframe belong. For progressive video, the video frame number is incremented sequentially. For interlaced video, the video frame number is incremented in steps of two. Therefore, each video frame has two frame numbers such that the video frames belonging to the first field have even frame numbers and the video frames belonging to the second field have odd frame numbers. Since the MAC header is coded using the most robust MCS, the error rate for the video header, which is included in the MAC header, is lower than the video subframe payload.

The PHY header always contains seven subframe headers irrespective of the availability of the corresponding subframe payloads. The length field indicates the length of the corresponding subframe payload. If a subframe payload is not present then the length field is set to zero. The HRP mode index field represents the MCS used for the subframe. The possible values available for the HRP mode index are shown in Table 5.5 in Section 5.3.4. The PHY control field contains an unequal error protection (UEP) field that together with the HRP mode index can be used to differentiate between UEP-by-mapping and UEP-by-coding modes. We will discuss UEP in more detail later.

The standard, low-latency and AV aggregation schemes are summarized in Table 9.1.

9.2.2.4 Block-ACK

For each subframe payload, the FCS may either carry a single cyclic redundancy checksum (CRC) computed over both least significant bits (LSBs) and most

Table 9.1 Summary of aggregation schemes in IEEE 802.15.3c (2009) [11]. Reproduced by permission © 2009 IEEE

	Standard	Low latency	Audio video
Individual subframes may coded separately using with different MCS	Yes	No	Yes
Maximum number of subframes aggregated in one MPDU	8	256	7
Efficient support of audio/video	No	No	Yes
Allows fragmentation of MSDUs carrying video data	Yes	Yes	No
Support bidirectional low latency data	No	Yes	No
Robust protection of video header	No	No	Yes

significant bits (MSBs) or two checksums computed separately over LSBs and MSBs of the subframe payload. The destination device indicates the status of the subframes in the previously received MAC frame by sending a block-ACK bitmap. For the two-checksum case, the nth MSB field indicates the status of the MSB portion of the nth subframe and the nth LSB field indicates the status of the LSB portion of the nth subframe. However, in the one-checksum case, one bit in the block-ACK represents the status of both MSB and LSB portions of a subframe.

9.3   Design Considerations for Supporting Uncompressed Video

In data communications, all bits are equally important; hence, they should be equally protected.[2] In contrast, in uncompressed video streams, some bits are more important than others. For instance, compared to the LSB, the MSB of a color pixel has the maximum impact on the video quality [13]. Therefore, bits can be treated differently and it is not always necessary to deliver all bits with the same error control scheme. UEP provides a way to protect bits in the order of their importance. The bit error rate (BER) for high-importance bits is much lower. Numerous studies in the past have shown the benefits of using UEP at the PHY in the context of compressed video [14]. In addition, uncompressed video streams contain rich spatial redundancy, which can be used to overcome some pixel errors.

We have developed a system using the IEEE 802.15.3c AV OFDM mode for supporting uncompressed video streaming over 60 GHz wireless networks, as shown in Figure 9.8 [1]. The application layer at the video source implements pixel partitioning such that pixels with minimal spatial distance (i.e. neighboring pixels) are placed in different video packets. If a video packet retransmission is corrupted, then the receiver recovers the error using pixel information in other received packets containing neighboring pixels. As a result, further retransmission of corrupted pixels is not required. The MAC layer uses the AV aggregation mode to aggregate multiple video packets into one MAC frame. For each video packet, the MAC layer supports two CRC fields: MSB and LSB CRCs.

At the physical layer, information bits are first scrambled to randomize the input sequence. Then the four MSBs are parsed into the first data path, and the second four LSBs are parsed into the second data path. On each data path, Reed–Solomon (RS) and convolutional codes are concatenated to protect the information bits. We consider RS code (224, 216, $t = 4$) with Hamming distance $d_{\min} = 2t + 1$ [15]. We assume a color depth (i.e. the number of bits per color component) of 8 bits. However, the proposed system can be easily extended to other video streams using a deeper color depth (i.e. 12- or 16-bit color).

---

[2]Portions of this section are taken from [1, 12]. Reproduced by permission © 2008, 2009 IEEE.

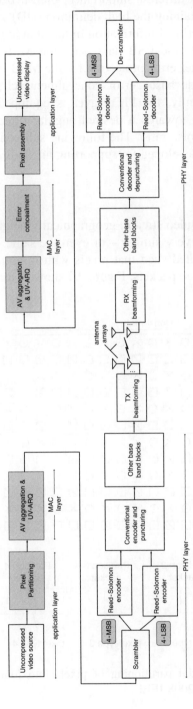

Figure 9.8 Block diagram of the transmitter and the receiver of the millimeter-wave system; shaded blocks are discussed in this chapter [1]. Reproduced by permission © 2008 IEEE.

The two bitstreams are of different importance; the MSB bitstream carries more critical information in maintaining the high definition (HD) video quality, while the LSB bitstream carries less critical information in maintaining the HD video quality. Therefore, in comparison to the LSB data path, the MSB data path is strongly protected, which allows better error protection for the MSB portion of video pixels. At the receiver, an RS code based error concealment scheme (RSS) is used to overcome pixel errors. Finally, the PHY layer is equipped with array antennas which allows beamforming towards a desired angular direction to maximize the signal-to-interference-and-noise ratio. The following subsections present a detailed description of the modules developed for supporting uncompressed video streaming.

9.3.1 Pixel Partitioning

In a typical uncompressed video stream, geographically neighboring (spatially correlated) pixels usually have very similar – or even the same – values. This kind of spatial redundancy is exploited such that pixels with minimal spatial distance are partitioned into different video packets. Figure 9.9 shows a diagrammatical example

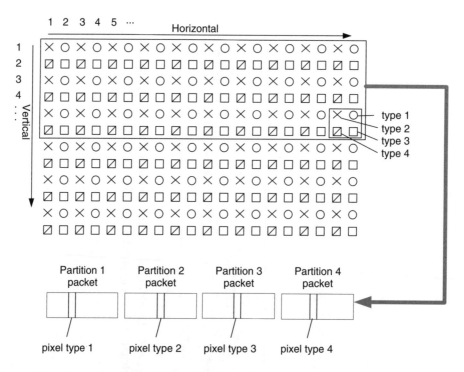

Figure 9.9 Example of spatial partitioning of pixels into four partition packets [1]. Reproduced by permission © 2008 IEEE.

of pixel partitioning and packetizing scheme in which four neighboring pixels are partitioned into four video packets. If one video packet is corrupted, then one or more other packets that contain pixels that are spatially related to the corrupted pixel(s) can be used to recover or compensate for the corrupted pixel information.

9.3.2 Uncompressed Video ARQ

Wireless systems designed for data communication send a packet with a checksum appended. The receiver recomputes the checksum. If an error is detected by failed checksum, the receiver requests the retransmission of the whole packet. It is possible that a chunk of bits are correctly received; however, the sender retransmits another copy of the whole packet.

The uncompressed video ARQ (UV-ARQ) protocol improves upon this by using two bits per video packet in the ACK to indicate the status of both MSB and LSB portions. If the MSB portion is received correctly, then the retransmission is not solicited. Otherwise, a robust modulation and coding mode (MSB only, as shown in Table 5.5) is used to reliably retransmit the MSB portion only. Figure 9.10 illustrates the functioning of the UV-ARQ, which can be summarized as follows:

- The sender appends multiple CRCs per video packet, and transmits the packet to the receiver.
- The receiver recomputes checksum for the MSB and LSB portions. The receiver signals to the sender concerning the status of the MSB and LSB portions.

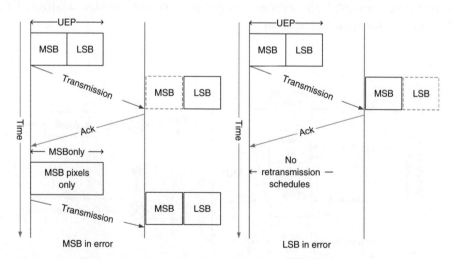

Figure 9.10 Illustration of UV-ARQ protocol sequence: if the LSB portion is received in error then no retransmission is scheduled [1]. Reproduced by permission © 2008 IEEE.

- If the MSB portion is correctly received then the sender skips the retransmission of the LSB portion as shown in Figure 9.10 (LSB error case). Otherwise, the sender retransmits the MSB portion only using the *MSB only* (MSB error case).
- Since the same *time* is granted to retransmit the MSB portion as for the original video packet, a reduced rate modulation index can strongly protect the retransmitted MSB portion against channel errors by proving a higher SNR.

9.3.3 Unequal Error Protection

UEP provides an efficient method to protect bits with different weighting. Figure 9.11 shows a generalized UEP structure [13]. UEP assumes multiple use of error control coding blocks, and applies different coding rates r_i to multiple input bitstreams. In fact, multi-Gbps transmission systems inevitably employ parallel signal processing blocks due to the limit of processing speed, and parallel or multiple error control coding blocks are natural in the PHY design.

- *UEP by coding*. A lower coding rate is allocated to more important bits (i.e. MSBs) and a higher coding rate to less important bits (i.e. LSBs) – for example, the MSBs (bits 7, 6, 5 and 4) with a lower coding rate than the LSBs (bits 3, 2, 1 and 0). UEP can also be provided by mapping in which some bits are strongly protected in comparison to other bits in the constellation diagram (see Section 5.3 for more discussion).
- *UEP by mapping*. Bits mapped onto the I-branch get stronger (unequal) protection than the bits on the Q-branch. Therefore, the constellation diagram looks like a rectangle; however, the average energy per symbol remains unaffected (see Section 5.3 for more discussion).

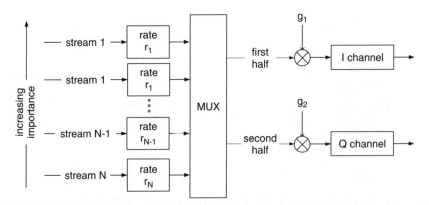

Figure 9.11   UEP by mapping and coding [1]. Reproduced by permission © 2008 IEEE.

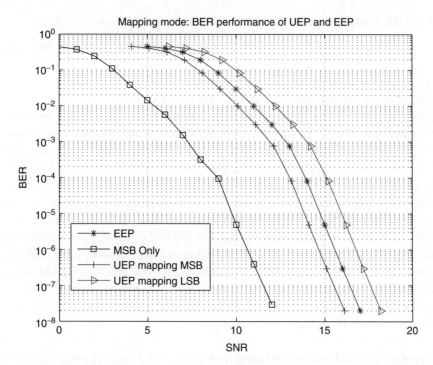

Figure 9.12 BER performance of EEP and UEP-by-mapping modes [1]. Reproduced by permission © 2008 IEEE.

In both UEP modes, the BER performance of MSBs is boosted at the expense of poor BER for LSBs since the MSB portions are strongly protected. The two separate CRCs for MSB and LSB portions help in limiting the error concealment to the erroneous portion of the video packet only. A correctly received portion of the video packet is forwarded to the higher layers as it is. Figure 9.12 presents the BER performance of UEP (by mapping) and equal error protection (EEP) modes of AV OFDM PHY. Notice that there is generally a difference of about 1.5–2 dB between the MSB performance and the EEP performance, and about 1–2 dB difference between the EEP performance and LSB performance.

9.3.4 Error Concealment

Uncompressed video requires huge bandwidth for retransmission, therefore an unlimited number of retries may not be desirable because of additional latency and buffer requirements at the receiver (i.e. display such as HDTV). We consider one-time retransmission of MSB portions. If the retransmitted packet is also corrupted, the receiver invokes an error concealment as presented in [12].

9.3.4.1   RS Code Swap (RSS) Error Concealment Scheme

We use RS codes to conceal some pixel errors. We combine *good* (i.e., uncorrupted) RS codes from a corrupted video packet and adjacent partitions to reconstruct the original video packet. While developing the RSS error concealment scheme, we consider a *cross-layer feedback* from the PHY to the MAC.

In the AV OFDM, each video packet consists of 100 RS codes to achieve a high channel efficiency, thus meeting the delay constraints of uncompressed video. Therefore, the length of each video packet is 21,600 bytes, and one $1920 \times 1080p$ (HD) frame is evenly divided into 288 video packets. The RS code $(224, 216, t = 4)$ considered in the 60 GHz system can correct errors of up to 4 symbols (bytes). If more than 4 symbols are in error, it flags up an uncorrectable codeword. We use this kind of *feedback* from the PHY. We consider a cross-layer feedback from the PHY to MAC such that for each video packet, the PHY (i.e. RS decoder) signals to the MAC layer those RS codewords received correctly and those in error. Afterwards, the MAC layer (or the application layer) conceals the effect of failed RS codes on the video quality. Identified failed RS codes are replaced with *good* RS codes having pixels with minimum spatial variations. For a video packet, if the receiver detects an error, it takes the following steps:

- Erroneous RS codewords are identified at the PHY and signaled to the MAC layer.
- RS codes at the same position in other video packets from the same partition are used to replace the erroneous RS code. As shown in Figure 9.13, RS code j in video packet 1 is received in error. One of the RS codes at the same position j, which carry neighboring pixels, from video packet 2, 3 or 4 is used to replace the faulty codeword.
- If the previous step could not be successfully completed because none of the three adjacent partitions had the same indexed RS code correctly received, one of the adjacent *good* RS codes within the corrupted packet (e.g. RS code $j + 1$ or RS code $j - 1$ in video packet 1) is used to replace the erroneous codeword. In the next step, adjacent RS codes from different partitions are used.
- Finally, if some of the codewords could not be concealed, then display them as they are.

9.4   Performance Study

In this section, the performance of the AV system developed in Section 9.3 is presented. The ns2-based IEEE 802.15.3 MAC simulator is enhanced by implementing the new features described in Section 9.3. The PHY layer supports both the UEP (by mapping) and EEP modes. We consider the peak signal-to-noise ratio (PSNR)

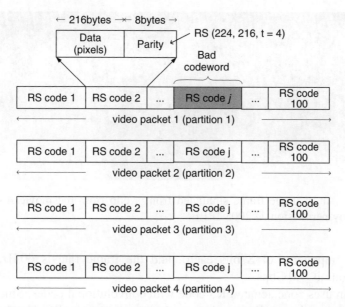

Figure 9.13 Illustration of RSS error concealment scheme. A video packet is composed of 100 RS codewords. RS code j in packet 1 is in error. Correctly received RS code j in other packets (2–4) from the same partition can be used to conceal the wrong code [1]. Reproduced by permission © 2008 IEEE.

as the key performance metric. For a received $N_1 \times N_2$ 8-bit image, the PSNR is given by

$$PSNR = 20 \log_{10} \left(\frac{255}{\sqrt{\frac{1}{N_1 N_2} \sum_{i=0}^{N_1-1} \sum_{j=0}^{N_2-1} [f(i, j) - F(i, j)]^2}} \right), \qquad (9.1)$$

where $f(i, j)$ is the pixel value of the source video frame, and $F(i, j)$ is the pixel value of the reconstructed video frame at the display. N_1 and N_2 are equal to 1080 and 1920, corresponding to the 1920×1080p video format, respectively. The measured PSNR indicates the difference between the transmitted and the received video frame. The average PSNR is defined as

$$\overline{PSNR} = \frac{1}{K} \sum_{i=1}^{K} PSNR_i, \qquad (9.2)$$

where K is the total number of uncompressed video frames simulated. We simulate 1000 from a clip from the movie *Alexander*; an example frame from the movie is shown in Figure 9.14. Each frame has 1920×1080 pixels, each pixel has 24 bits

Figure 9.14 An example frame simulated from a clip from the movie *Alexander* [1]. Reproduced by permission © 2008 IEEE.

(i.e. RGB components of 8 bits each), and the frame rate is 60 Hz. Thus, the application rate is 3.0 Gbps.

The system uses concatenated RS code with convolutional codes. Since the errors at the Viterbi decoder are bursty, they tend to present to the RS decoder correlated symbol errors. From [15] we get the following relation between codeword error probability (P_w) and bit error probability (P_b):

$$P_b \approx \frac{d_{\min}}{n} P_w, \tag{9.3}$$

where $d_{\min} = 9$ and $n = 224$ given the RS code (224, 216, $t = 4$). In the event of error, $d_{\min}$ bytes in a codeword are randomly flipped. We evaluate the performance of the 60 GHz system, presented in the previous section, under random uniform errors. In the simulation study, video data are coded either UEP or EEP. In both cases, we consider no retransmissions and resort to the RSS scheme to conceal pixel errors. While performing the simulation study, we made realistic assumptions such that given the application rate of 3 Gbps and transmission rate of 3.807 Gbps (MCS-2 and MCS-4 shown in Table 5.5), approximately 20% extra bandwidth is available. We assume MAC and other processing overhead of 10%, and therefore the remaining extra bandwidth is used to retransmit only 10% of the total transmitted packets without exceeding the timing requirements of video signals. One-time limited retransmissions can further improve the quality of video signals at the expense of additional buffer and processing overhead.

9.4.1 Effect of UEP and EEP

The EEP mode treats the MSB and LSB portions of a video packet equally. The MSB portions, which contribute more to the PSNR, are strongly protected in the

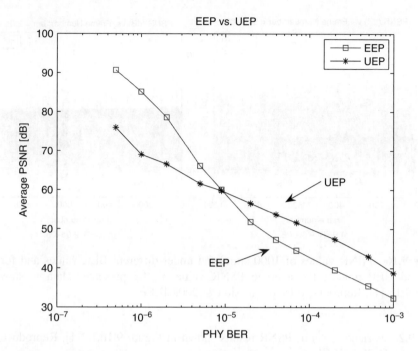

Figure 9.15 Average PSNR values for the EEP and UEP modes [1]. Reproduced by permission © 2008 IEEE.

UEP mode; however, the LSB portions are weakly protected. This is because in the low BER range (BER $< 9 \times 10^{-6}$), the average PSNR of the EEP mode outperforms UEP (see Figure 9.15). However, in the high BER range (BER $> 9 \times 10^{-6}$), UEP achieves better average PSNR because MSB portions are strongly protected thereby maintaining a high PSNR (see Figure 9.15). For the BER values we simulated in this study, the UEP mode always maintains PSNR values greater than 40 dB which is generally accepted as good picture quality. An adaptive EEP/UEP scheme, with EEP at low BER and UEP at high BER, is also feasible, but UEP alone can maintain a good picture quality over the entire BER range.

9.4.2   Stability of UEP

Figure 9.16 presents the PSNR values of one thousand frames simulated for the EEP and UEP modes. The corresponding average PSNR values are shown in Figure 9.15. Table 9.2 summarizes the variance of PSNR values shown in Figure 9.15. Notice that the variances of PSNR values for the UEP mode are much smaller than the corresponding results for the EEP mode. This suggests that UEP results in less

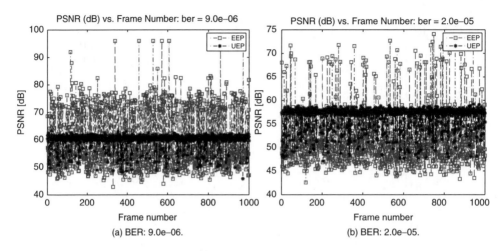

Figure 9.16 PSNR values of 1000 simulated under different BER values and for both EEP and UEP modes; the average PSNR value of the presented data is shown in Figure 9.15 [1]. Reproduced by permission © 2008 IEEE.

Table 9.2 Variance (σ^2) of PSNR results shown in Figure 9.16(b) [1]. Reproduced by permission © 2008 IEEE

PHY BER	2.0×10^{-6}	5.0×10^{-6}	9.0×10^{-6}	2.0×10^{-5}	4.0×10^{-5}	7.0×10^{-5}
EEP	245.51	196.32	112.78	32.95	10.89	9.58
UEP	6.95	4.19	5.93	5.75	5.90	5.21

fluctuating PSNR values than EEP. Although in some cases the EEP mode attains a higher mean PSNR value, the stability effect of the UEP mode provides much better visual quality than the EEP mode because, for most human observers, wide fluctuations in the picture quality result in more severe visual degradation.

9.4.3 VQM Scores

The video quality metric (VQM) [16] is an objective metric that computes the magnitude of the visible difference between the transmitted and received video sequences, and larger visible degradations result in larger VQM values. For instance, a zero value represents no impairment between the original and processed video sequence. VQM scores for UEP are lower than for EEP in the high BER range as shown in Figure 9.17. In Figure 9.15 we observed similar behavior with the PSNR

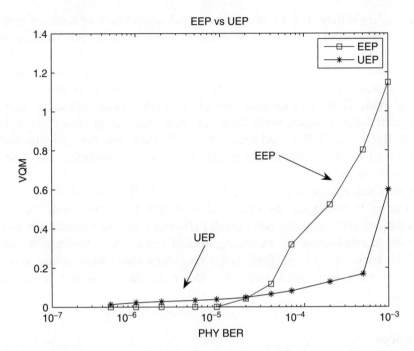

Figure 9.17 Average VQM scores for the EEP and UEP modes [1]. Reproduced by permission © 2008 IEEE.

metric; however, the UEP/EEP cross-over point for VQM has slightly shifted from 9.0×10^{-6} to 2.0×10^{-5}.

9.5 Conclusions and Future Directions

With the availability of low-cost 60 GHz CMOS circuits, there is no doubt that a large number of new devices will soon be available that can support emerging applications such as high-quality uncompressed video streaming and fast large file transfers (such as photos, high-definition images, audio/video content) over short distances. However, 60 GHz directional transmission creates many difficulties at the MAC layer. In the recent past, researchers have proposed many MAC solutions in the context of IEEE 802.11 that rely on a combination of directional and omnidirectional transmissions. However, in 60 GHz it is difficult to realize omnidirectional transmissions. Therefore, a TDMA-like protocol is more suitable at 60 GHz.

Compared to low-frequency signals (2.4 or 5 GHz), 60 GHz signals incur significantly higher path losses. However, at shorter distances (less than 10 meters), 60 GHz can support multi-gigabit data rates. Therefore, the two wireless bands provide widely different, yet complementary, characteristics in terms of range and throughput. For these reasons, the viability of multi-band WLAN devices that are capable of operating in both the lower-frequency band (e.g. 2.4 and/or 5 GHz) and higher-frequency band (60 GHz) becomes very attractive. In this way, multi-band stations achieve the best of both worlds by combining the coverage of the lower band and multi-Gbps transmission rate of the higher band. However, there is a greater need for an efficient multi-band operation that enables fast session transfer from 60 GHz to 2.4/5 GHz, and vice versa [17], such that the optimum interface is used based on the communicating stations' physical proximity and multi-band channel conditions.

Another challenge is to increase the range of 60 GHz transmissions. Advanced beamforming technologies, as discussed in Chapter 4, can surely help. MAC layer solutions such as multi-band mesh networks can also enhance the coverage of 60 GHz transmissions by exploiting spatial reuse and coordination from the 2.4/5 GHz band. Yang and Park [18] have presented some initial results on the benefits multi-band mesh networks. However, more research is required on this topic.

References

[1] Singh, H., Oh, J., Kweon, C., Qin, X., Shao, H. and Ngo, C. (2008) A 60 GHz wireless network for enabling uncompressed video communication. *IEEE Communications Magazine* **46**(12), 71–78.

[2] Ko, Y.-B., Shankarkumar, V. and Vaidya, N. (2000) Medium access control protocols using directional antennas in ad hoc networks. *Proceedings of IEEE Infocom*.

[3] Nasipuri, A., Ye, S., You, J. and Hiromoto, R. (2000) A MAC protocol for mobile ad hoc networks using directional antennas. *Proceedings of IEEE Wireless Communications and Networking Conference* (WCNC).

[4] IEEE 802.11 (2007) IEEE Standard Part 11: Wireless LAN Medium Access Control (MAC) and Physical Layer (PHY) Specifications.

[5] Takai, M., Martin, J., Bagrodia, R. and Ren, A. (2003) Directional virtual carrier sensing for directional antennas in mobile ad hoc networks. *Proceedings of 3rd ACM International Symposium on Mobile Ad Hoc Networking & Computing* (MobiHoc-2003), pp. 183–193.

[6] Choudhury, R., Yang, X., Ramanathan, R. and Vaidya, N. (2002) Using directional antennas for medium access control in ad hoc networks. *Proceedings of 8th Annual International Conference on Mobile Computing and Networking* (MobiCom-2002), pp. 59–70.

[7] ElBatt, T., Anderson, T. and Ryu, B. (2003) Performance evaluation of multiple access protocols for ad hoc networks using directional antennas. *Proceedings of IEEE Wireless Communications and Networking Conference (WCNC)*, pp. 982–987.

[8] Korakis, T., Jakllari, G. and Tassiulas, L. (2003) A MAC protocol for full exploitation of directional antennas in ad-hoc wireless networks. *Proceedings of 3rd ACM International Symposium on Mobile Ad Hoc Networking & Computing* (MobiHoc-2003), pp. 98–107.

[9] Choudhury, R. and Vaidya, N. (2004) Deafness: a MAC problem in ad hoc networks when using directional antennas. *Proceedings of IEEE International Conference on Network Protocols* (ICNP), pp. 283–292.

[10] IEEE 802.15.3b (2005) Part 15.3: Wireless Medium Access Control (MAC) and Physical Layer (PHY) Specifications for High Rate Wireless Personal Area Networks (WPANs), Amendment 1: MAC Sublayer.

[11] IEEE P802.15.3c/D15 (2009) Part 15.3: Wireless Medium Access Control (MAC) and Physical Layer (PHY) Specifications for High Rate Wireless Personal Area Networks (WPANs): Amendment 2: Millimeter-Wave Based Alternative Physical Layer Extension, July.

[12] Singh, H., Niu, H., Qin, X., Shao, H., Kweon, C., Fan, G., Kim, S. and Ngo, C. (2008) Supporting uncompressed HD video streaming without retransmissions over 60 GHz wireless networks. *Proceedings of IEEE Wireless Communications and Networking Conference (WCNC)*, pp. 1939–1944.

[13] Kim, S. et al. (2007) UEP for 802.15.3c PHY. IEEE 802.15-07/701r4, July.

[14] Zheng, W., Tan, Y.-P., Liu, B. and Yiin, L.-H. (1996) A novel approach to unequal error protection for image/video delivery with source detector. *Proceedings of SPIE: Video Techniques and Software for Full-Service Networks*.

[15] Benedetto, S. and Biglier, E. (1999) *Principles of Digital Transmission: with Wireless Applications*. New York: Kluwer Academic/Plenum Press.

[16] Pinson, M. and Wolf, S. (2004) A new standardized method for objectively measuring video quality. *IEEE Transactions on Broadcasting* **50**(3), 312–322.

[17] IEEE 802.11ad (2009) Very high throughput in 60 GHz. http://www.ieee802.org/11/Reports/tgad_update.htm

[18] Yang, L. and Park, M. (2008) Applications and challenges of multi-band gigabit mesh networks. *Proceedings of SENSORCOMM*, pp. 813–818.

10

Remaining Challenges and Future Directions

Alberto Valdes-Garcia, Pengfei Xia, Su-Khiong Yong and Harkirat Singh

This book has presented a thorough overview of fundamental elements of 60 GHz technology, covering key layers of the communication protocol stack, from channel modeling to medium access control. Current research and development results have also been presented to illustrate the evolving nature of this technology. With two completed standards, multiple industry consortia and the first products to enter the market, 60 GHz has firmly established itself as the next major trend in gigabit wireless technology. Nevertheless, as is apparent from the contents of this book, new challenges have arisen at every layer of the communication protocol, which require a significant degree of innovation to be overcome. Moreover, beyond technical advancements at particular layers, a successful 60 GHz solution (technically and economically) will need to address all aspects of the system design simultaneously.

Some of the most important challenges remaining and anticipated developments are as follows.

Full system evaluation in end-user scenarios. As seen in Chapters 7 and 8, the implementation of a full 60 GHz demonstrator from radio to baseband is a challenging task from many perspectives. Despite the availability of such prototypes, to perform full system-level evaluations and measurements over different circumstances (channel conditions, reflections, interferences, etc.) is also a difficult task. As a result, experimental data on the performance of 60 GHz

60 GHz Technology for Gbps WLAN and WPAN: From Theory to Practice
Edited by Su-Khiong (SK) Yong, Pengfei Xia and Alberto Valdes Garcia
© 2011 John Wiley & Sons, Ltd

systems under realistic usage scenarios is still limited. For this reason, some fundamental questions related to system design might not have a definitive answer and are subject to endless debate based on simulation results, limited experiments and extrapolation of results available from other technologies. The upcoming commercial deployment of beam-steered, Gbps, wireless solutions is expected to provide means to evaluate different beamforming algorithms, modulation formats, MAC strategies, etc. The relative performance of different solutions in real-life situations (in terms of their cost, form factor, latency and throughput) may have to await large-scale commercial deployment.

Compensation and robust design for PHY non-idealities. The attainment of data throughputs in the Gbps range comes at the price not only of bandwidth but also high precision in the processing of the carrier modulated signal, namely very low phase noise, high IQ phase and amplitude balance, high amplifier linearity, ADC and DAC accuracy. Some of these challenges were described quantitatively in Chapter 3, and they will become harder to overcome as the demand for greater data rate increases. In this context, digital compensation and calibration techniques for RF non-idealities will become more crucial as technology nodes proceed deeper down in the nanometer regime. Modulation and coding techniques that are inherently robust to these physical layer limitations will also provide a competitive advantage.

System-in-package integration and testing. The cost of packaging and testing is expected to dominate over the cost of IC fabrication for high-volume 60 GHz applications. Therefore, the development of design-for-testability techniques, built-in-self-test, and known-good die techniques at millimeter-wave frequencies will become an important part of product development. Recent results show that the package-level integration of individual silicon beamforming modules with small numbers of elements may be feasible for modular and low-cost 60 GHz beam-steered solutions [1]. Planning for system partition into one or several integrated circuits will go hand-in-hand with packaging concepts (SoC, SiP) and testing strategies at each level of production and integration. The most cost-effective solutions will come from the right balance among these three aspects for a particular application and volume. A separate and equally important issue is the need of methodologies for standard-compliance testing methods. These exist for wireless technologies at lower carrier frequencies and helped to create a testing and manufacturing environment that reduced costs and accelerated product development.

MAC design for higher throughput and power efficiency. The MAC layer has several challenges in supporting application data rates greater than 10 Gbps. A multi-channel directional MAC would be a natural choice in these scenarios. Efficient control of directional multiple channels in terms of device discovery and resource allocation would then become necessary. Such scenarios would also demand the development of more intelligent hybrid ARQs than exist today. Moreover, to facilitate the use of 60 GHz technology in portable devices, the MAC layer needs to help extend the battery life. This can be achieved not only by developing smart power management but also by reducing overhead in device discovery, beamforming, association and data transmission. The next generation 60 GHz MAC should consider these open issues.

Higher network throughput. Most of the current research and design efforts at 60 GHz focus on individual point-to-point links. As 60 GHz devices gain wider adoption, the 60 GHz frequency band may become congested. The support of multiple transmissions at the same time (e.g. how to improve the overall network throughput) must be addressed. Fortunately, we can take advantage of the fact that most 60 GHz transmissions are highly directional by using spatial division multiple access (SDMA) [2]. Multiple directional 60 GHz point-to-point links could potentially coexist with each other, if the interference between links could be kept under control. This calls for a significant research effort in the area of multi-user MIMO, interference suppression, interference alignment, and the developments will go hand-in-hand with those in the MAC layer as discussed in the previous paragraph.

Challenges in MIMO channel modeling. A 60 GHz MIMO solution is critical to our ability to deliver the promise of multi-gigabit wireless solutions. Hence, realistic 60 GHz MIMO channel models based on virtual antennas and/or multiple-antenna systems measurement are required for all the typical environments in which 60 GHz technology is to be deployed. This allows communications designers to estimate the achievable data rates and diversity gain in the 60 GHz MIMO channels and produce a robust system design. For example, the characterization of transmit and receive antenna correlations, which serve as key MIMO design parameters, is lacking at 60 GHz. This information gap persists despite the fact that large antenna arrays are expected to be used in this band. Similarly, a multiple dual-polarized MIMO channel model is also lacking and deserve more investigation. While these models are needed for future 60 GHz systems, the complexity and cost involved in performing such measurement are high if not prohibitive.

The already existing 60 GHz technical approaches described in this book, as well as others that will emerge to surmount the above-mentioned challenges, will create a rich knowledge base from which other millimeter-wave Gbps technologies (not necessarily limited to 60 GHz frequency bands) and systems will emerge. Here are some examples of future directions.

Higher-data-rate 60 GHz systems. The natural trend for the evolution of current systems is to achieve even higher data rates. One of the main drivers for this trend is the ever increasing data rate requirements of high-definition video. According to the WiHD Next Generation specification announced recently [3], up to 28 Gbps may be required to support applications of this kind. If the current channelization is to be kept unchanged, data rates higher than 10 Gbps may only be possible by channel bonding, or the use of simultaneous parallel links.

Long-range point-to-point links. In the frequency range of 65–115 GHz (approximately) the long-range oxygen absorption of millimeter-wave energy is significantly smaller (~10 dB) than at 60 GHz. This has motivated the use of frequency bands at 71–76 and 81–86 GHz (E-band) and even 100 GHz for kilometer-range Gbps point-to-point links [4, 5]. Commercial systems in these bands formally existed before 60 GHz commercial products, not as integrated solutions but rather based on monolithic microwave integrated circuit modules with III-V semiconductors. Although the market volume and channel characteristics for these systems will remain significantly different from those at 60 GHz, they will be affected by the current developments on silicon integration of millimeter-wave radios and Gbps digial modem modules. One application for E-band systems currently under consideration is de-congesting backhaul connections in cellular networks [6, 7].

Wireless networking in data centers and digital systems. Data centers are the backbone of information technology infrastructure worldwide. In recent years, a large increase in the demand for their data-processing capacity, and the requirement to moderate the energy consumption, among other important factors, have exerted pressure to rethink their architecture. Efficient networking is critical to the performance of these massive clusters of servers and, until recently, increasing the bandwidth of existing optical and electrical connections was the only option. However, the availability of Gbps wireless links has motivated the consideration of other interesting alternatives in the way data centers of the future may be realized [8, 9]. Highly dense digital systems of smaller footprint than data centers (i.e. systems in package) face networking challenges that are equally daunting. In these cases millimeter-wave propagation is also being explored as an alternative to hard-wired connections at distances of a few millimeters [10].

In years to come, what is now regarded as an emerging technology will be remembered as just the starting point of a series of major innovations in the way we design, produce and utilize wireless communications.

References

[1] Wambacq, P., Raczkowski, K., Ramon, V., Vasylchenko, A., Enayati, A., Libois, M. et al. (2009) Low-cost CMOS-based receive modules for 60 GHz wireless communication. *IEEE Compound Semiconductor Integrated Circuits Symposium*, pp. 1–4, October.

[2] Xia, P., Yong, S. K., Ngo, C. and Oh, J. (2008) A practical SDMA protocol for 60 GHz millimeter wave communications. *42nd Asilomar Conference on Signals, Systems and Computers*, October.

[3] Wireless HD Next Generation Specification Announcement, http://www.wirelesshd.org/pdfs/WiHD%20Next%20Gen%20Jan10%20FINAL.pdf

[4] Wells, J. (2009) Faster than fiber: The future of multi-Gbs wireless. *IEEE Microwave Magazine*, **10**(3), 104–112.

[5] Kosugui, T., Hirata, A., Nagatsuma, T. and Kado, Y. (2009) MM-wave long-range wireless systems. *IEEE Microwave Magazine*, **10**(2), 68–76.

[6] Chia, S., Gasparroni, M. and Brick, P. (2009) The next challenge for cellular networks: backhaul. *IEEE Microwave Magazine*, **10**(5), 54–66.

[7] Lockie, D. and Peck, D. (2009) High-data-rate millimiter-wave radios. *IEEE Microwave Magazine*, **10**(5), 75–83.

[8] Ramachadran, K., Kokku, R., Mahindra, R. and Rangarajan, S. (2008) 60 GHz data-center networking: wireless worry less? NEC Technical Report, July.

[9] Kandula, S., Padhye, J. and Bahi, P. (2009) Flyways to de-congest data center networks. *Eighth ACM Workshop on Hot Topics in Networks*, October.

[10] Kawasaki, K. et al. (2010) A millimeter-wave intra-connect solution, *IEEE International Solid-State Circuits Conference*, pp. 414–415, February.

Index

60 GHz Technology for Gbps WLAN and WPAN: From Theory to Practice
Edited by Su-Khiong (SK) Yong, Pengfei Xia and Alberto Valdes Garcia
© 2011 John Wiley & Sons, Ltd